Leckie

the education publisher
for Scotland

Second Level

MATHS

for S1-S3

3

Author: Sheona Goodall
Series Consultant: Carol Lyon
Series Editor: Craig Lowther

Contents

Answers

Check your answers to this workbook online: https://collins.co.uk/pages/scottish-primary-maths

1.1 Rounding whole numbers

1 Write the multiple of 10 000 that comes before and after each number.

Draw an ↓ showing where you think the actual number lies on the number line.
Round each number to the nearest 10 000.

a) 37 252 lies between [30 000] and [40 000] rounds to [ ]

b) 637 252 lies between [ ] and [ ] rounds to [ ]

c) 657 252 lies between [ ] and [ ] rounds to [ ]

d) 35 049 lies between [ ] and [ ] rounds to [ ]

2 Round these numbers to the nearest 100 000. Draw an empty number line to help.

a) 426 209 rounded [ ]

b) 456 209 rounded [ ]

c) 556 209 rounded [ ]

d) 624 195 rounded [ ]

e) 724 195 rounded [ ]

f) 784 500 rounded [ ]

3 Scott and Willow have been rounding numbers. They notice that they have written different answers for some of the questions. Look at their answers and decide which are correct and which are incorrect.

	Number to round	Scott	Willow
a)	284 967	280 000	300 000
b)	52 914	50 000	50 000
c)	773 820	770 000	770 000
d)	350 906	300 000	300 000

Write the correct answers into the boxes below and tick the name of the student you agree with.

Scott　　Willow

a) 284 967 rounded to the nearest 100 000 ☐　☐　☐

b) 52 914 rounded to the nearest 10 000 ☐　☐　☐

c) 773 820 rounded to the nearest 10 000 ☐　☐　☐

d) 350 906 rounded to the nearest 100 000 ☐　☐　☐

★ Challenge

1. Use these clues to find the correct number from the table. Circle your answer.

446 371	386 421	320 465	366 251
403 862	488 231	43 284	411 840

- I am a six-digit number.
- I have exactly 4 even digits, but I am odd.
- When you round me to the nearest 10 000, I round up.
- I am 400 000 when rounded to the nearest 100 000.

2. Use these clues to find the correct number from the table. Circle your answer.

852 185	78 145	819 445	749 315
793 604	776 825	802 218	830 560

- When you round me to the nearest 10 000, I round down.
- I am a multiple of 5.
- I am 800 000 when rounded to the nearest 100 000.

1.2 Rounding decimal fractions

1 Round each decimal number to the nearest hundredth:

a) 4·732 rounds to

b) 4·734 rounds to

c) 4·737 rounds to

d) 4·735 rounds to

e) 4·7316 rounds to

f) 74·738 rounds to

2 Circle True or False for each number statement.
Write the correct answer for those that are false.

a) 5·623 rounded to the nearest hundredth is 5·63   **True   False**

b) 5·693 rounded to the nearest hundredth is 5·69   **True   False**

c) 5·616 rounded to the nearest hundredth is 5·62   **True   False**

d) 25·185 rounded to the nearest hundredth is 25·19   **True   False**

e) 25·183 rounded to the nearest hundredth is 25·180 **True   False**

3 Round these measurements to two decimal places:

a) 2·059 litres [   ]

b) 81·488 kg [   ]

c) 0·073 cm [   ]

d) 418·571 km [   ]

★ Challenge

1. Circle the numbers that give 37·43 when rounded to the nearest hundredth. One has been done for you.

   37·421   37·435   37·4218   37·4302   (37·429)   37·4306

2. Find as many six-digit decimal numbers as you can that give 8·29 when rounded to the nearest hundredth.

1.3 Using rounding to estimate the answer

1 Use rounding and estimating to help you to decide if each answer is reasonable and explain. The first one has been done for you.

a) $17\,428 - 4283 = 13\,145$

(Reasonable)	Not Reasonable	Rounding to the nearest 1000, this is $17\,000 - 4000$. $13\,000$ is a good estimate so $13\,145$ is reasonable.

b) $36 \times 18 = 928$

Reasonable	Not Reasonable	

c) $24\,740 + 5429 = 30\,169$

Reasonable	Not Reasonable	

d) $72\,375 - 23\,862 = 48\,513$

Reasonable	Not Reasonable	

e) $45 \times 21 = 9540$

Reasonable	Not Reasonable	

f) $275\,812 - 122\,401 = 153\,411$

Reasonable	Not Reasonable	

2 Say whether each answer is reasonable or not. Explain your answer.

	Reasonable		Explain your answer
	Yes	No	
Zahid multiplied 48 by 9 and got 432.			
Alana multiplied 79 by 6 and got 513.			
Marek multiplied 93 by 7 and got 651.			

⭐ **Challenge**

Use estimation to help you find three numbers from those shown that total approximately 100 000.

Do you think there is a better solution? Try a different set of three numbers.

68 416   71 829
25 980
9817   81 904
38 055
17 356   4203

2.1 Reading and writing whole numbers

1 Write these six-digit numbers in words.

a) 143 284

b) 143 519

c) 143 020

d) 743 020

e) 753 020

f) 700 020

2 Write these six-digit numbers in numerals.

a) eight hundred and twenty-five thousand, three hundred and seventy-one

b) eight hundred and five thousand, three hundred and seventy-one

c) two hundred and five thousand, three hundred and seventy-one

d) five hundred and five thousand, three hundred and seventy-one

e) five hundred thousand and thirteen

f) seven hundred and twenty-two thousand, nine hundred

3 This is the key for a secret code:

0	1	2	3	4	5	6	7	8	9
@	$	%	&	*	§	±	{	~	^

Work out the six-digit number. Write it in numerals and then in words.

a) | & | § | ± | { | @ | ^ |

b) | % | ~ | $ | ± | * | % |

c) | § | @ | & | ~ | & | @ |

★ **Challenge**

6430102

In words, this number says **six million, four hundred and thirty thousand, one hundred and two**.

Write the number that is two million more than this number in both words and numerals.

2.2 Representing and describing whole numbers

1 Here is the number 214 336 made with place value arrow cards.

| 2 0 0 0 0 0 ▷ | 1 0 0 0 0 ▷ | 4 0 0 0 ▷ | 3 0 0 ▷ | 3 0 ▷ | 6 ▷ |

| 2 1 4 3 3 6 ▷ |

Here is the same number made with place value counters.

100 000   10 000   1000   1000   100   10   1   1   1

100 000   1000   1000   100   100   10   10   1   1   1

Write each number here in numerals. You could use place value arrow cards **or** place value counters to help you.

a) three hundred and twenty thousand, two hundred and fourteen

b) three hundred thousand, two hundred and fourteen

c) five hundred thousand, two hundred and fourteen

d) five hundred thousand, two hundred and four

2 For each number, write the value of the underlined digit in both words and numerals. The first one has been done for you.

a) 5<u>6</u>2 804   **Sixty thousand, 60 000**

b) <u>5</u>62 804

c) 562 80<u>4</u>

12

d) <u>3</u>18 241

e) 318 2<u>8</u>1

f) 318 2<u>4</u>1

g) 3<u>1</u>8 241

h) 9<u>6</u>0 377

3

| 4 | 7 | 0 | 1 | 8 | 5 |

Use the digits on these cards to find a number to match each clue.
Write your answer in words and numerals.

a) Find an even six-digit number with zero in the thousands place.

b) Find the smallest six-digit number where the value of the 5 is 50.

⭐ **Challenge**

Find the smallest number you can make using **all** these cards.
Write your number in both words and numerals.

four   and   fifty

thousand   hundred   million

eighteen   three

2.3 Place value partitioning of whole numbers

1 The number 631 825 has been partitioned in six different ways (and more partitions are possible!)

600 000 + 30 000 + 1000 + 800 + 20 + 5

631 000 + 800 + 20 + 5

630 000 + 1000 + 800 + 20 + 5

631 825

631 000 + 825

600 000 + 31 825

630 000 + 1825

Find six different ways to partition each of these numbers.

641 825

541 825

541 625

2 Write the number represented by these place value houses in four different ways. The first one has been done for you.

a)

Thousands			Ones		
H	T	O	H	T	O
4	2	3	8	5	9

423 thousands, 8 hundreds, 5 tens and 9 ones.
423 thousands, 8 hundreds, 59 ones.
423 thousands, 85 tens and 9 ones.
423 thousands, 859 ones.

b)

Thousands			Ones		
H	T	O	H	T	O
4	2	3	2	5	9

c)

Thousands			Ones		
H	T	O	H	T	O
1	2	3	2	5	9

d)

Thousands			Ones		
H	T	O	H	T	O
1	2	3	2	4	9

★ Challenge

Anna is partitioning the number 725 844. She says:

I can partition this in two ways. It can be 725 thousands, 8 hundreds and 44 ones. It can also be 700 thousands, 258 hundreds and 4 ones.

Do you agree? Explain how you know.

2.4 Number sequences

1 Write the next five numbers in each sequence.

a) 528 286, 528 287, 528 288,

[   ] , [   ] , [   ] , [   ] , [   ]

b) 528 296, 528 297, 528 298,

[   ] , [   ] , [   ] , [   ] , [   ]

c) 528 996, 528 997, 528 998,

[   ] , [   ] , [   ] , [   ] , [   ]

d) 529 993, 529 994, 529 995,

[   ] , [   ] , [   ] , [   ] , [   ]

e) 549 997, 549 998, 549 999,

[   ] , [   ] , [   ] , [   ] , [   ]

f) 299 996, 299 997, 299 998,

[   ] , [   ] , [   ] , [   ] , [   ]

2 Now write the next five numbers in each of these sequences.

a) 712 314, 712 313, 712 312,

[   ] , [   ] , [   ] , [   ] , [   ]

b) 712 404, 712 403, 712 402,

[   ] , [   ] , [   ] , [   ] , [   ]

c) 712 006, 712 005, 712 004,

[   ] , [   ] , [   ] , [   ] , [   ]

d) 780 006, 780 005, 780 004,

	,		,		,		,	

e) 800 006, 800 005, 800 004,

	,		,		,		,	

f) 400 002, 400 001, 400 000,

	,		,		,		,	

3 Look at these number sequences. Are the numbers increasing or decreasing? How much bigger or smaller are they getting each time? One has been done for you.

a) 370 004, 470 004, 570 004, 670 004

The numbers are increasing. They get bigger by 100 000 each time.

b) 413 501, 413 401, 413 301, 413 201

c) 175 850, 174 850, 173 850, 172 850

d) 599 988, 599 998, 600 008, 600 018

Adam and Lynda made two number sequences. In their sequences they used the numbers on each of these cards only once. One sequence had four numbers in it and the other had five numbers in it. In one sequence the numbers were increasing and in the other sequence the numbers were decreasing. Write down what you think the two sequences were.

2.5 Comparing and ordering whole numbers

1 Write < or > in each box to make each statement true.

a) 420 381 ☐ 420 821

b) 200 720 ☐ 200 702

c) 634 100 ☐ 63 410

d) 508 650 ☐ 508 655

2 Circle True or False for each of these statements. Change some words in the false statements to make them true. Cross out the word you are changing and write the new word in the answer box.

a) 403 200 = four hundred and three thousand, two hundred   **True**   **False**

☐

b) 311 131 < three hundred thousand one hundred and eleven   **True**   **False**

☐

c) 602 544 > sixty-two thousand five hundred and forty-four   **True**   **False**

☐

d) 100 010 < ten thousand and ten   **True**   **False**

☐

3 Write each set of numbers in ascending order (smallest to largest).

a) 523 162   523 149   523 154   523 160   523 151

b) 187 960   187 951   187 966   187 909   187 941

c) 638 194   638 199   638 180   638 189   638 191

4 a) Use the numerals on the cards to make different six-digit numbers that fit the criteria.

0   5   4   9   3   6

- A multiple of 5

- An odd number

- The smallest possible number

- A multiple of 10

- An even number

- A multiple of 25

b) Now put your numbers in descending order (largest to smallest).

1. Use these cards to make nine different six-digit numbers.

Four hundred and seventeen thousand...	...three hundred and twenty-five
Two hundred and forty-one thousand...	...three hundred and fifty-two
Four hundred and forty-one thousand...	...two hundred and eighteen

2. Write all nine numbers in ascending order (smallest to largest).

2.6 Negative numbers

1 These tables show the midday temperature on 1st December in some European cities.

Athens	12°C
Brussels	2°C
Cologne	−1°C
Edinburgh	3°C

Helsinki	−5°C
Istanbul	7°C
Kyiv	−4°C
Lisbon	11°C

Oslo	−3°C
Paris	0°C
Rome	5°C
Zurich	−2°C

1. In which cities was the midday temperature below freezing?

2. Which city had the coldest midday temperature on this day?

3. How many degrees warmer was it in Rome than in Oslo?

4. Athens was 14°C warmer than which city?

5. What was the difference in temperature between Lisbon and Kyiv?

6. What was the difference in temperature between the coldest and the warmest places?

7. Write the names of the cities in order, from coldest to warmest.

A diver uses an underwater drone to take some photos of a shipwreck. The depth that each photo was taken at is shown here:

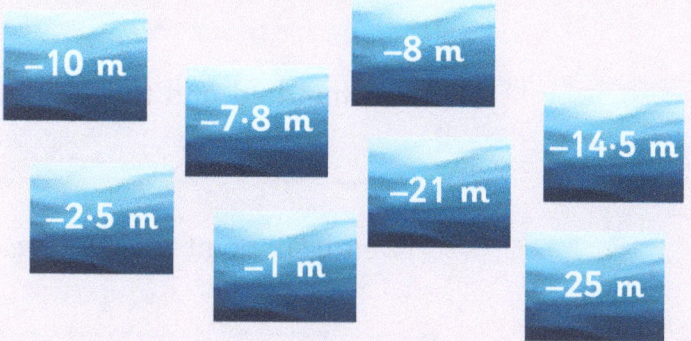

Draw a picture of the underwater scene and label it with these depths in the correct order.

1. Match each of these decimal fractions with the correct speech bubble. One has been done for you.

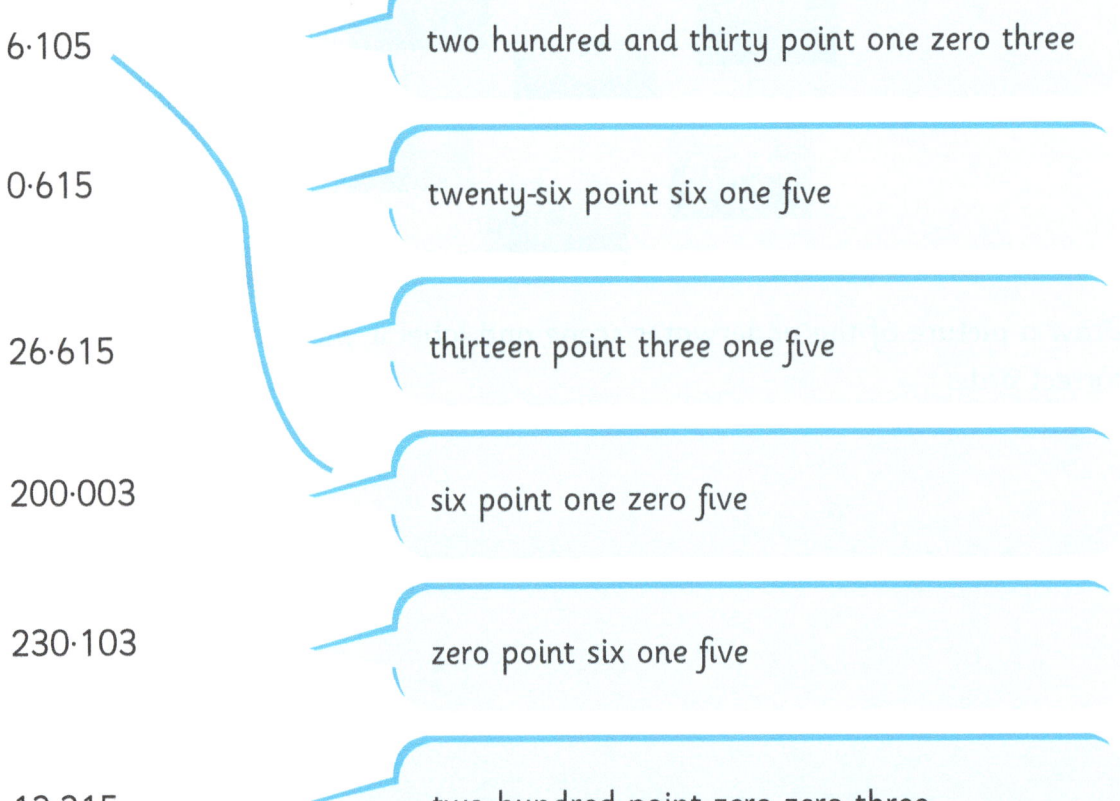

6·105

0·615

26·615

200·003

230·103

13·315

two hundred and thirty point one zero three

twenty-six point six one five

thirteen point three one five

six point one zero five

zero point six one five

two hundred point zero zero three

2. 4 wholes and 183 thousandths can be written as 4·183

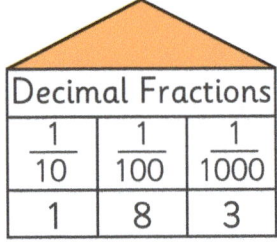

Ones				Decimal Fractions		
H	T	O	•	$\frac{1}{10}$	$\frac{1}{100}$	$\frac{1}{1000}$
		4		1	8	3

Write in numerals:

a) 7 wholes and 351 thousandths

b) 7 wholes and 391 thousandths

c) 27 wholes and 391 thousandths

d) 27 wholes and 396 thousandths

e) 52 wholes and 184 thousandths

f) 0 wholes and 485 thousandths

g) 257 wholes and 5 thousandths

h) 9 wholes and 46 thousandths

★ **Challenge**

A conifer in the school grounds is 4.950 metres high. Lilya says "That is 4 metres and 950 thousandths of a metre". Sami says "No, it is 4 metres and 95 thousandths of a metre". Who do you think is correct? Explain.
You can use the place value houses to help you.

2.8 Representing and describing decimal fractions

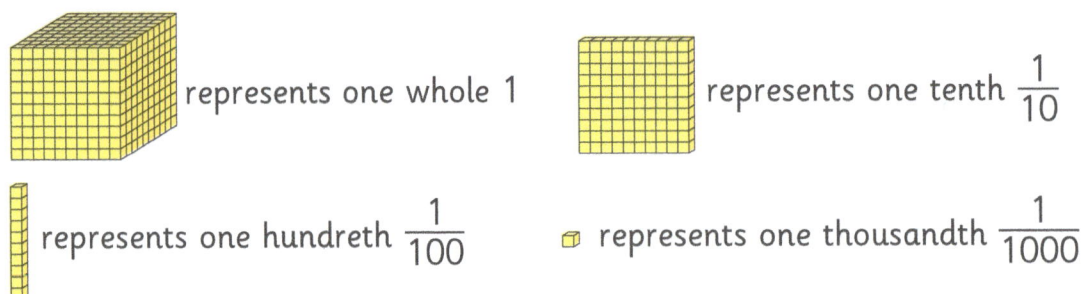

represents one whole 1

represents one tenth $\frac{1}{10}$

represents one hundreth $\frac{1}{100}$

represents one thousandth $\frac{1}{1000}$

This model represents the decimal fraction 2·416

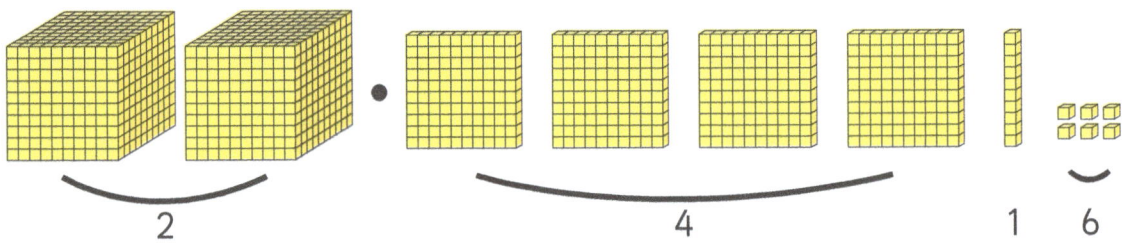

2   4   1   6

We can write this decimal fraction in three ways, like this:

2 ones, 4 tenths, 1 hundredth and 6 thousandths = 2·416 = $2\frac{416}{1000}$

---

1   Write the decimal fraction represented by the following models in three ways.

a)

b)

c)

d)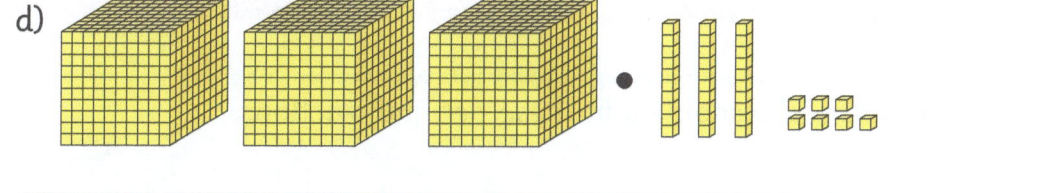

e)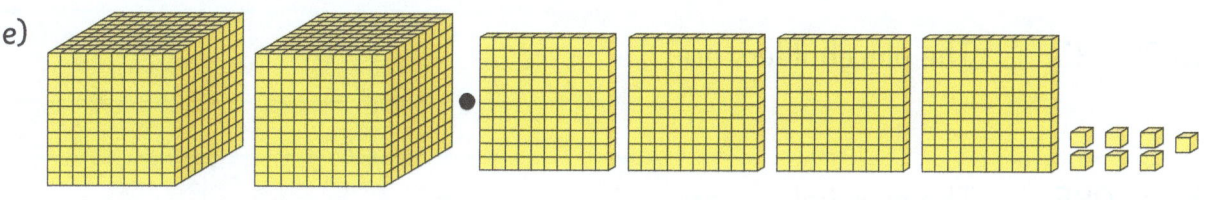

f)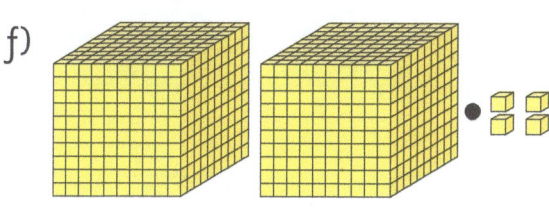

★ **Challenge**

Some students built a model of a decimal fraction with thousandths and then wrote the decimal fraction in three different ways. They had to break the model down and when they looked at their answers some paint had been spilled on the page. Can you work out what the decimal fraction was?

3 thousands = █ · 2 █ = 5 $\frac{█}{1000}$

We will need █

█ enths

9 hu █

█ sands

2.9 Zero as a placeholder in decimal fractions

1 Describe the position of the placeholders in these decimal fractions.
For example:

23·004 → the placeholders are in the tenths and hundredths places

a) 30·185   The placeholders are

b) 30·085   The placeholders are

c) 30·005   The placeholders are

d) 296·107   The placeholders are

e) 206·007   The placeholders are

f) 801·530   The placeholders are

2 This diagram represents 1 whole and 37 hundredths.
We can write this as a fraction or as a decimal fraction.

$1\frac{37}{100}$ *one and thirty-seven hundredths*

1·37 *one point three seven*

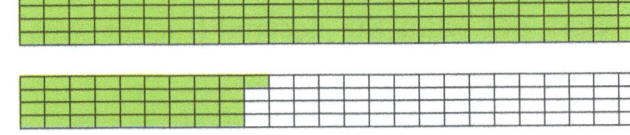

Circle True or False for each of these. Rewrite the false statements to make them true.

a) $7·39 = 7\frac{39}{100}$ **True**   **False**

b) $7·09 = 7\frac{9}{1000}$ **True**   **False**

c) $7·095 = 7\frac{95}{1000}$ **True**   **False**

d) $15 \cdot 8 = 15\frac{8}{10}$ **True**   **False**

e) $30 \cdot 06 = 30\frac{6}{100}$ **True**   **False**

f) $41 \cdot 106 = 41\frac{16}{100}$ **True**   **False**

3   Using some or maybe all of the grids provided, draw a diagram to show each of these.

a) $0 \cdot 3 = 0 \cdot 30$

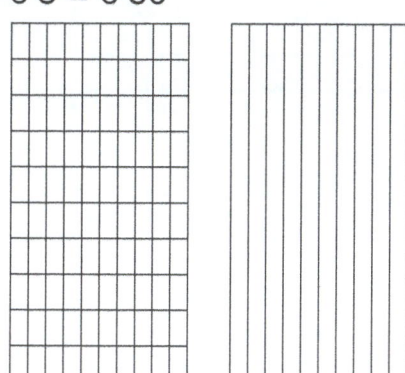

b) $0 \cdot 3 + 0 \cdot 04 = 0 \cdot 34$

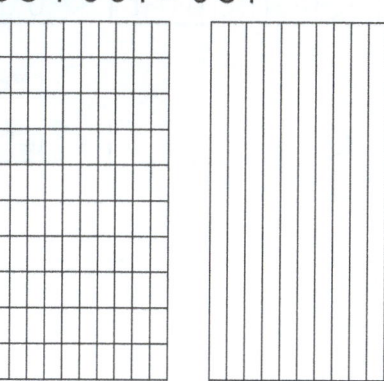

★ Challenge

1. Find as many pairs of equal numbers as you can. One has been done for you.

10·050

0·4

1·05

2·70

2·7

16·10

10·05

$1\frac{5}{100}$

0·04

5·201

$16\frac{1}{10}$

2. For the numbers without a pair, write a matching number on the diagram and join them up.

2.10 Partitioning decimal fractions

1 Write the decimal fraction that can be made using these place value arrow cards.

a) `3·` `0·6` `0·0 1` `0·0 0 9`

b) `5·` `0·6` `0·0 1` `0·0 0 9`

c) `5·` `0·2` `0·0 1` `0·0 0 9`

d) `5·` `0·2` `0·0 1` `0·0 0 4`

e) `0·` `0·8` `0·0 7` `0·0 0 3`

g) `0·0` `0·0 4` `0·0 0 2`

2 Partition these decimal fractions. For example: **4·65 = 4 + 0·6 + 0·05**

a) 9·2

b) 9·7

c) 9·1

d) 7·61

e) 7·68

f) 7·08

g) 2·164

h) 2·144

i) 2·044

j) 6·005

★ Challenge

1. Match the place value counters to the correct decimal fraction.

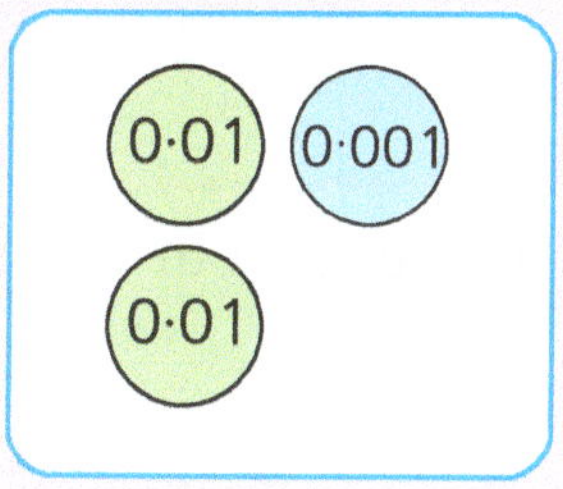

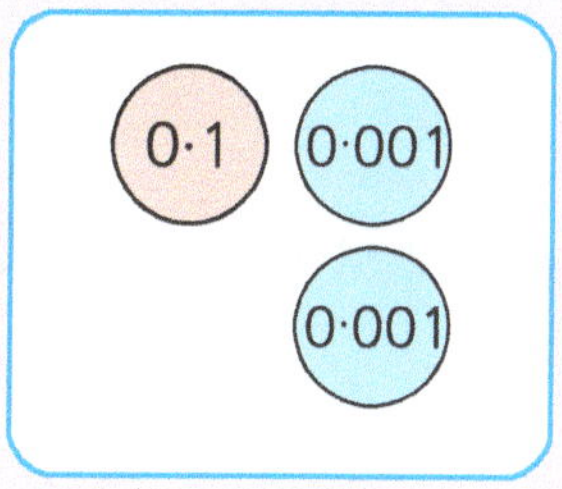

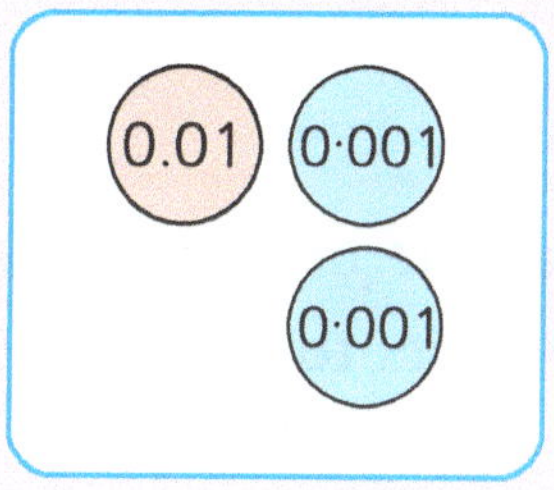

| 0·01 0·001 | 0·1 0·001 | 0·01 0·001 |
| 0·01 | 0·001 | 0·001 |

| 0·102 | 0·012 | 0·021 |

2. Now make up three examples of your own like this.

2.11 Comparing and ordering decimal fractions

1 Write these decimal fractions in order from smallest to largest.

a) 2·518   7·518   0·518   5·518   4·518

b) 6·273   6·573   6·073   6·773   6·173

c) 5·428   5·421   5·425   5·420   5·426

d) 0·754   0·479   0·27   0·081   0·603

2 Circle True or False for each of these statements. Change the symbol in the false statements to make them true.

a) 2·73 > 2·6   **True**   **False**

b) 16·4 = 16·40   **True**   **False**

c) 5·07 = 5·7   **True**   **False**

d) 8·3 < 8·14   **True**   **False**

e) 9·02 > 9·11   **True**   **False**

f) 7 = 7·000   **True**   **False**

3 Complete each of the following statements. The first one has been done for you.

a) 6 hundredths and 3 thousandths = [63] thousandths = [0.063]

b) 4 hundredths and 9 thousandths = [   ] thousandths = [   ]

c) 5 tenths and 8 hundredths and 0 thousandths = 58 [   ] = [   ]

d) 300 thousands = [   ] hundredths = [   ] tenths = 0.3

⭐ **Challenge**

Five athletes take part in a long jump event at an athletics competition. The lengths of their jumps are as follows:

3·094 m   3·1 m   3·64 m   3·09 m   3·52 m

1. How far was the longest jump? [   ]

2. List the jumps in order, starting with the smallest.

[   ]

3. Lexi takes part in the event after these five athletes have jumped. She makes the second longest jump. What is the longest she could have jumped and the shortest she could have jumped?

[   ]

3.1 Mental addition and subtraction

1 Calculate using a mental strategy of your choice.

a) 13 478 + 999

b) 17 478 + 999

c) 17 478 + 1999

d) 26 300 + 491

e) 26 300 + 891

f) 31 278 + 3005

g) 16 600 − 398

h) 16 600 − 598

i) 16 600 − 1998

j) 42 765 − 304

k) 21 648 − 1009

l) 50 831 − 7002

Which mental strategy did you use most when doing these calculations? Explain your thinking below. If you can, talk to a partner or an adult about your strategy.

2 Use place value and number facts to calculate each of these.

a) $104\,502 + 70\,000$

b) $104\,502 + 20\,000$

c) $304\,502 + 20\,040$

d) $701\,430 + 130\,021$

e) $678\,302 - 30\,000$

f) $678\,302 - 50\,000$

g) $293\,804 - 120\,200$

h) $580\,273 - 320\,001$

★ **Challenge**

1. What number is ten thousand more than $499\,900$?

2. In 2020, the population of Glasgow was $632\,350$ and the population of Edinburgh was $506\,520$.

   a) What was the total population of both cities in 2020?

   b) How many more people were living in Glasgow than in Edinburgh at that time?

3.2 Adding and subtracting a string of numbers

1 Add these strings of numbers.

a) 347 + 53 + 12 + 8000 + 2000 + 426

b) 347 + 53 + 12 + 9000 + 2000 + 426

c) 447 + 53 + 12 + 9000 + 2000 + 426

d) 2300 + 70 + 730 + 1700 + 391 + 29

e) 2300 + 60 + 740 + 1700 + 391 + 29

f) 2700 + 60 + 740 + 1300 + 391 + 29

g) 185 + 1445 + 68 + 632 + 2555 + 15

2 Find the missing digits in each of these calculations and fill them in to complete the number sentences.

a) 3273 + 1 ☐ ☐ ☐ + 607 = 4920

b) 7 ☐ ☐ ☐ − 420 − 380 = 6560

3 Subtract each set of numbers from the starting number in bold.

a) Start with **16 380**. Subtract 280, 316 and 154 from this number.

b) Start with **14 270**. Subtract 280, 316 and 154 from this number.

c) Start with **14 270**. Subtract 290, 326 and 144 from this number.

d) Start with **27 000**. Subtract 2038, 460 and 562 from this number.

e) Start with **35 050**. Subtract 2100, 85, 1900 and 115 from this number.

★ Challenge

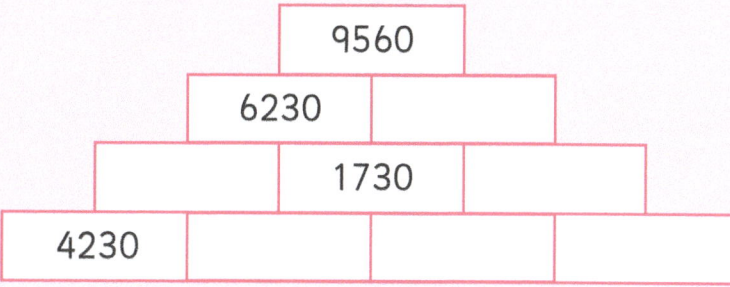

In this number pyramid, the number in each block is the total of the two numbers in the blocks below it. Find all the missing numbers.

3.3 Using place value partitioning to add and subtract

1. We can use an empty number line and partitioning into thousands, hundreds, tens and ones to help us add and subtract six-digit numbers. For example, to find:

37 560 + 240 315

I could start with 240 315 and partition 37 560 into 30 000 + 7000 + 500 + 60

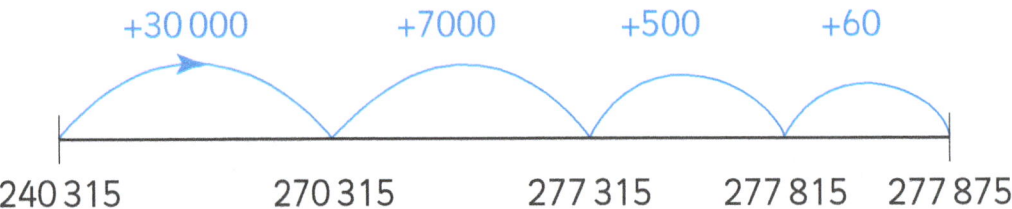

37 560 + 240 315 = 277 815

Use an empty number line and partitioning to calculate:

a) 108 463 + 47 315

b) 25 374 + 53 481

c) 108 463 + 43 315

d) 108 463 + 63 315

e) 59 230 + 380 610

f) 83 564 − 32 173

g) 47 249 − 23 136

h) 145 628 − 32 173

i) 145 628 − 91 329

j) 473 525 − 418 361

2 We can also use the column method to help us add and subtract six-digit numbers.
For example:

$736\,423 - 424\,306$

```
   736 423
 − 424 306
 ─────────
   300 000
    10 000
     2 000
       100
        20
 −       3
 ─────────
   312 117
```

Use this method to calculate:

a) $25\,674 + 23\,481$

b) $162\,481 + 241\,537$

c) $147\,249 − 113\,255$

d) $635\,816 − 521\,809$

3 Choose a suitable method to work out each of these.

a) 492 517 + 186 374

b) 863 795 − 704 518

★ Challenge

	755 000	370 000	623 040	581 309	180 260	630 000	245 000	376 960	

Match pairs of cards that total one million. If any of the cards do not have a match, write the matching number on one of the blank cards.

3.4 Adding whole numbers using standard algorithms

1 We can use a standard written algorithm to help us with addition calculations that are too tricky to work out mentally. For example:

35 416 + 17 836 + 28 533

```
    2 1   1
  3 5  4 1 6
  1 7  8 3 6
+ 2 8  5 3 3
  ─────────
  8 1  7 8 5
```

Use a standard written algorithm to work out the answer to each of these additions.

a) 47 108 + 38 265

b) 47 168 + 38 295

c) 47 568 + 38 895

d) 637 184 + 44 296

e) 637 184 + 144 296

f) 284 377 + 510 384

g) 294 377 + 540 384

h) 437 113 + 351 782

i) 437 113 + 391 782

2 Calculate the answers to these questions using the standard written algorithm for addition.

a) 41 267 + 37 482 + 12 866

b) 41 767 + 37 582 + 12 466

c) 71 767 + 47 582 + 32 466

d) 8462 + 175 355 + 36 714

3 These written algorithms contain some errors. Identify the errors then rewrite each algorithm correctly.

a)
```
   1 1 1
   1 7 4 8 2
 +   4 6 9 1
 ─────────────
   1 1 0 7 3
```

b)
```
         1
   2 3 4 8 3
 + 2 5 6 7 6
 ─────────────
   4 8 1 1 5 9
```

c)
```
       1 1
   1 2 9 3
 + 3 0 4 7 1
 ─────────────
   4 3 4 0 1
```

1. Fill in the missing digits in each of these written algorithms.

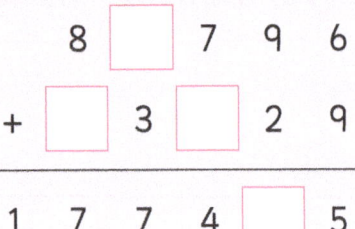

$$
\begin{array}{r}
8\ \boxed{3}\ 7\ 9\ 6 \\
+\ \boxed{9}\ 3\ \boxed{6}\ 2\ 9 \\
\hline
1\ 7\ 7\ 4\ \boxed{2}\ 5 \\
\end{array}
\qquad
\begin{array}{r}
8\ 9\ 5\ 6\ 3 \\
+\ 4\ \boxed{3}\ 5\ \boxed{5}\ 9 \\
\hline
\boxed{1}\ \boxed{3}\ 3\ \boxed{1}\ 2\ 2 \\
\end{array}
$$

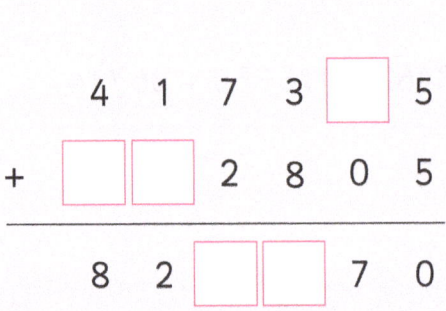

$$
\begin{array}{r}
4\ 1\ 7\ 3\ \boxed{6}\ 5 \\
+\ \boxed{4}\ \boxed{0}\ 2\ 8\ 0\ 5 \\
\hline
8\ 2\ \boxed{0}\ \boxed{1}\ 7\ 0 \\
\end{array}
\qquad
\begin{array}{r}
\boxed{2}\ 6\ 6\ \boxed{3}\ 4\ 8 \\
5\ 4\ \boxed{6}\ 9\ 0\ 1 \\
+\ 6\ \boxed{6}\ 4\ 5\ 9\ 0 \\
\hline
\boxed{1}\ 4\ 7\ 7\ 8\ \boxed{3}\ 9 \\
\end{array}
$$

2. Make up some missing digit addition algorithms of your own for a partner to solve.

3.5 Subtracting whole numbers using standard algorithms

1 We can use a standard written algorithm to help us with subtraction calculations that are too tricky to work out mentally. For example:

48 217 − 29 632

$$
\begin{array}{r}
3\ ^{17}\ ^{11}\ ^{1} \\
\cancel{4}\ \cancel{8}\ \cancel{2}\ 1\ 7 \\
-\ 2\ 9\ 6\ 3\ 2 \\
\hline
1\ 8\ 5\ 8\ 5 \\
\hline
\end{array}
$$

Use a standard written algorithm to work out the answer to each of these subtractions.

a) 53 824 − 16 532

b) 53 824 − 16 537

c) 653 824 − 446 537

d) 623 824 − 446 537

e) 180 743 − 29 588

f) 473 120 − 38 202

g) 728 541 − 534 146

h) 803 400 − 256 315

i) 901 612 − 529 745

2 Now, use a standard written algorithm to work out the answer to each of these subtractions. Check your answers by adding, like this:

48 217 – 29 632

```
   3 17 11 1
   4 8 2 1 7
 – 2 9 6 3 2          CHECK
   1 8 5 8 5    Add
```

```
   1 1 1
   2 9 6 3 2
 + 1 8 5 8 5
   4 8 2 1 7
```

a) 75 253 – 45 377

b) 75 053 – 45 877

c) 275 053 – 45 874

d) 215 053 – 145 874

e) 317 950 – 56 271

f) 317 950 – 156 271

3 These written algorithms contain some errors. Identify the errors then rewrite each algorithm correctly.

a)
```
   4 5 2 9 6
 - 1 8 4 0 8
 ───────────
   3 3 2 9 2
```

b)
```
      0 ¹²¹³ ¹
   ₁ ₃ ₄ 1 7 3
 -   8 5 2 0 1
 ─────────────
     4 6 9 7 2
```

c)
```
      0 ¹⁵¹⁵ ¹
   2 ₁ ₇ ₆ 0 4
 - 1 0 8 8 4 1
 ─────────────
   1 0 7 7 6 3
```

⭐ **Challenge**

1. Fill in the missing digits in each of these written algorithms.

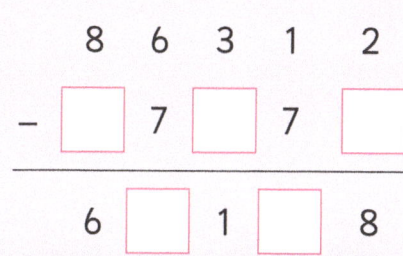
```
   8 6 3 1 2
 -  ☐ 7 ☐ 7 ☐
 ─────────────
   6 ☐ 1 ☐ 8
```

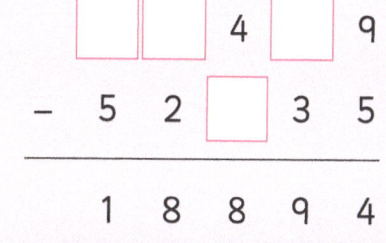
```
   ☐ ☐ 4 ☐ 9
 -   5 2 ☐ 3 5
 ─────────────
   1 8 8 9 4
```

2. Can you make up a missing digits subtraction of your own using five-digit numbers?

3.6 Mental and written calculation strategies

1 Calculate the following, choosing an efficient strategy for each one. Explain how you worked each answer out. Was it: partitioning, empty number line, mental calculation, algorithm, grouping or another strategy?

a) 130 000 + 270 000

b) 243 000 + 151 998

c) 430 000 + 10 001

d) 724 173 + 263 392

e) 398 745 + 400 020

f) 6284 + 3819 + 4537

2 Find the missing number on each brick in this subtraction pyramid by calculating the difference between the two numbers directly below it.

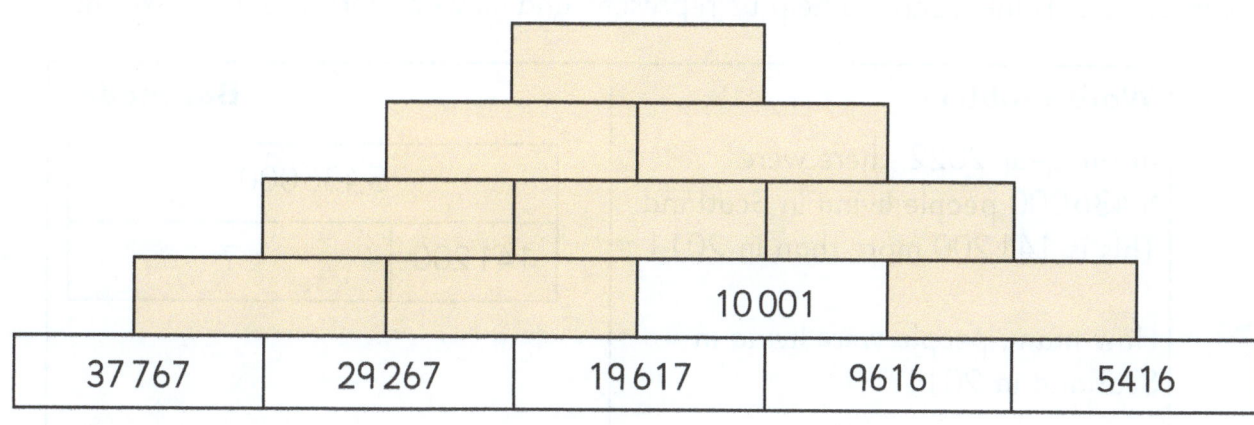

| 37 767 | 29 267 | 19 617 | 9616 | 5416 |

★ **Challenge**

1. Fill in the missing digits in each of these calculations.

   a) 3 ☐ 4 ☐ 1 + 1 800 = 34 20 ☐

   b) ☐ 7 ☐ 6 ☐ − 11 500 = 8 ☐ 963

2. Now create a missing digits subtraction calculation of your own using two five-digit numbers.

3.7 Representing word problems

We can draw a Think Board to help us represent and solve a word problem, like this:

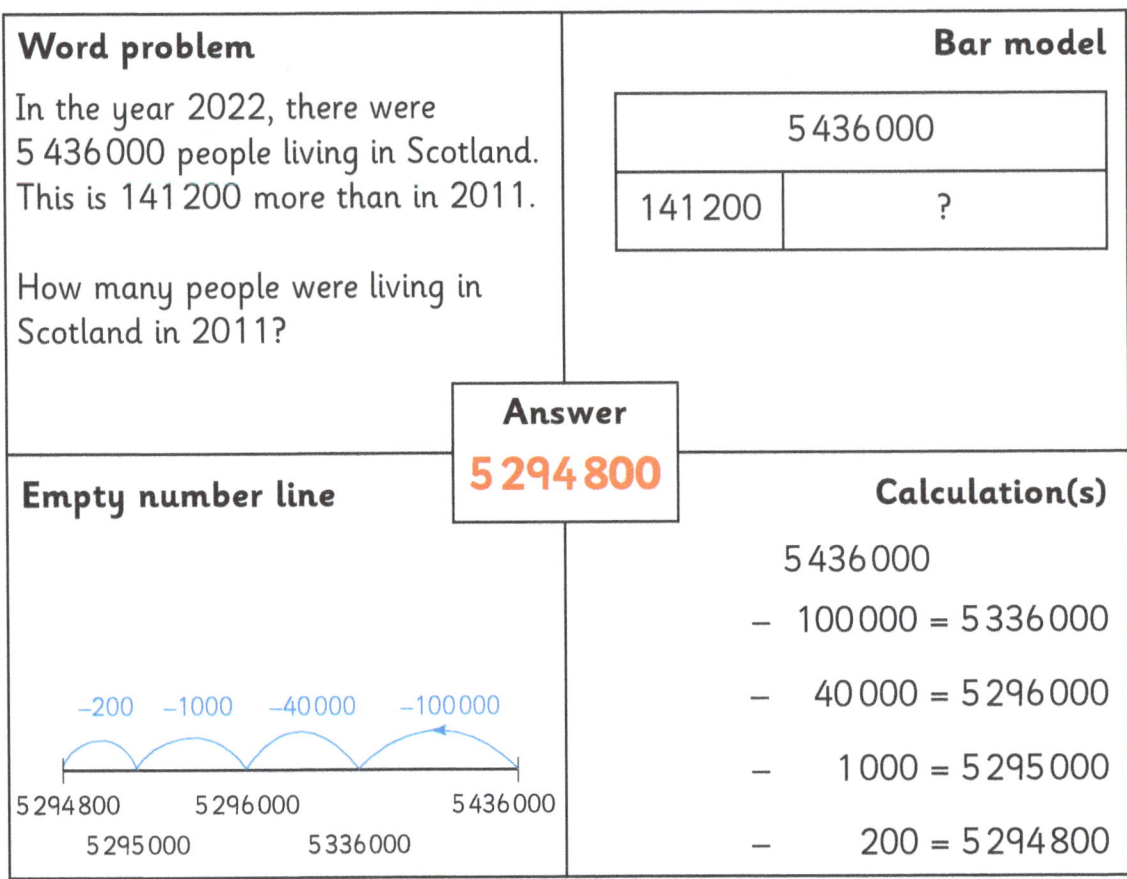

Word problem

In the year 2022, there were 5 436 000 people living in Scotland. This is 141 200 more than in 2011.

How many people were living in Scotland in 2011?

Bar model

5 436 000	
141 200	?

Answer

5 294 800

Empty number line

Calculation(s)

5 436 000

− 100 000 = 5 336 000

− 40 000 = 5 296 000

− 1 000 = 5 295 000

− 200 = 5 294 800

Complete the following Think Boards and solve each of these word problems.

1

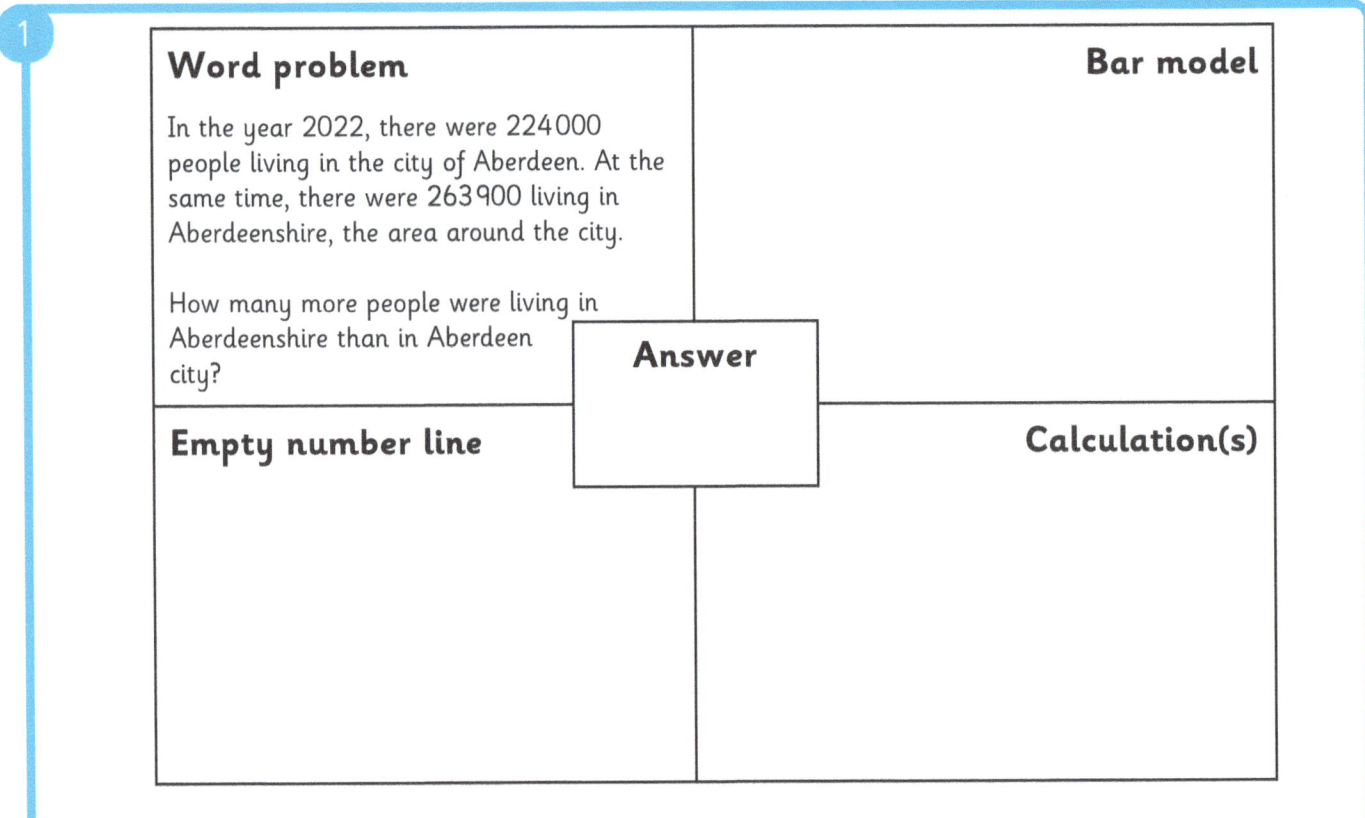

Word problem

In the year 2022, there were 224 000 people living in the city of Aberdeen. At the same time, there were 263 900 living in Aberdeenshire, the area around the city.

How many more people were living in Aberdeenshire than in Aberdeen city?

Bar model

Answer

Empty number line

Calculation(s)

2

Word problem

In 2022, Cardiff was the largest city in Wales with a population of 485 340 and Swansea was the second largest city with a population of 381 300.
How many people altogether were living in Cardiff and Swansea at this time.

Bar model

Answer

Empty number line

Calculation(s)

3

Word problem

In January, February and March of 2022, 255 000 people visited Scotland. In the next three months of 2022, the number of visitors was 520 000.

How many people visited Scotland in the first six months of 2022?

Bar model

Answer

Empty number line

Calculation(s)

4

Word problem	Bar model
In 2022, there were 832 300 people aged 15 and under living in Scotland. At the same time, 1 091 000 people over the age of 65 were living in Scotland. How many more over-65s were living in Scotland at that time?	

Answer

Empty number line	Calculation(s)

⭐ **Challenge**

Write a word problem for this partially completed Think Board. Complete the Think Board for your word problem.

Word problem	Bar model
	342 600 28 400

Answer

Empty number line	Calculation(s)

3.8 Solving multi-step word problems

Solve each of these problems by breaking it down into steps.

1. A local animal rescue charity was given £164 300 to put towards building an animal shelter. A local business donated an additional £25 000 towards building the animal shelter. The cost of building the shelter is £208 500. How much more money will the animal rescue charity need to raise before the shelter can be built?

2. In one year, a theme park attracted 237 600 visitors. A nearby science centre had 34 870 visitors. Half of the science centre visitors said they had also gone to the theme park. How many visitors went to the theme park **only**?

3. 1000 visitors to a sea life centre were asked what their favourite creature was. $\frac{1}{4}$ chose otters, 430 chose sea lions, 185 chose turtles and the rest said sharks. How many of the 1000 visitors chose sharks as their favourite creature?

4 Shay, Cian, Olive and Willow are doing a step-count challenge as part of Health Week. The target for their team of four is to complete 400 000 steps in a week. They keep a note of their step count for the first six days of the challenge:

	Total after six days
Shay	88 407
Cian	82 310
Olive	86 526
Willow	87 521

a) How many steps does the team still have to complete to meet their target of 400 000?

b) On the final day of the challenge Cian completes 12 453 steps and Willow completes 14 467 steps. Shay and Olive decide that they will each do half of the remaining steps to get the team to 400 000. How many steps will Shay and Olive each need to complete?

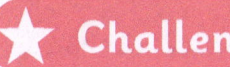
This table shows the population of some European countries:

Country	Population
France	64 200 000
Germany	83 310 000
Italy	58 870 000
Spain	47 520 000
United Kingdom	67 500 000

Write the answer to each of these problems in numerals and words.

1. How many more people live in the United Kingdom than live in France?

2. How much lower is the population of Spain than of Italy?

3. What is the difference between the most populated and the least populated countries in the table?

3.9 Adding whole numbers and decimal fractions

1 We can **round** to the nearest ten to help us when we are adding whole numbers and decimal fractions. For example:

5·81 + 27 is the same as 2·81 + 30   **The answer is 32·81**

Use this strategy to calculate:

a) 49 + 7·53

b) 39 + 7·53

c) 39 + 4·53

d) 8·71 + 146

e) 8·93 + 146

f) 8·93 + 149

g) 27 + 14·62

h) 127 + 14·62

i) 258 + 18·78

j) 43·91 + 308

k) 56·33 + 419

l) 837 + 89·19

2 We can **round** to the nearest whole number and **compensate** to help us when we are adding decimal fractions. For example:

5·96 + 7·61 is the same as 6 + 7·57   **The answer is 13·57**

Use this strategy to calculate:

a) 6·97 + 2·56

b) 6·94 + 2·56

c) 8·94 + 2·56

d) 8·94 + 5·81

e) 8·94 + 5·91

f) 5·27 + 1·99

g) 5·27 + 3·97

h) 9·36 + 9·94

i) 3·99 + 7·99

j) 6·96 + 6·96

k) 0·85 + 0·99

3 We can **partition** each decimal fraction into whole numbers and hundredths to help us when we are adding decimal fractions. For example:

14·62 + 6·15 = 14 + 6 + 0·62 + 0·15 = 20 + 0·77 = 20·77

Use this strategy to calculate:

a) 15·23 + 8·41

b) 15·23 + 8·36

c) 17·23 + 8·36

d) 36·18 + 20·64

e) 36·44 + 20·17

f) 60·37 + 25·37

4 Calculate each of these using a mental strategy of your choice.

a) 3·97 + 6·82

b) 19 + 5·39

c) 45·17 + 27·31

d) 28·31 + 36

e) 9·41 + 7·98

f) 19·38 + 29·51

1. Find as many pairs of numbers as you can in this grid that add to make 2460·53. Join these up with a line. One has been done for you.

560·52	310·23	1060·42
2400	1400·11	1900·01
2150·30	1000·03	60·53

2. Are there any numbers that you have not used? If there are, work out what the matching number would be to make a pair that gives a total of 2460·53.

3.10 Adding decimal fractions using standard written algorithms

1 We can use a standard written algorithm to help us add decimal fractions that are too tricky to work out mentally. For example:

$245 \cdot 76 + 383 \cdot 59$

```
  1   1   1
  2 4 5 · 7 6
+ 3 8 3 · 5 9
─────────────
  6 2 9 · 3 5
─────────────
```

Use a standard written algorithm to find the answers to these additions.

a) 37·48 + 15·31

b) 37·88 + 15·31

c) 37·88 + 15·79

d) 163·74 + 48·68

e) 163·92 + 48·68

f) 17·08 + 318·93

g) 17·08 + 358·93

h) 62·17 + 118·66

i) 58·85 + 460·37

2 Calculate the answers to these questions using the standard algorithm for addition.

a) 7·53 + 28·49 + 1·67

b) 12·77 + 5·38 + 0·89

c) 23·61 + 18·78 + 38·97

d) 59·72 + 63·18 + 86·39

⭐ **Challenge**

Fill in the missing digits in each of these addition calculations.

1.
```
    □ 4 · □ 6
+ 2 □ · 8 □
─────────────
  6 4 · 8 3
─────────────
```

2.
```
    □ 2 · 9 □
  4 □ · 8 4
+ 3 7 · □ 9
─────────────
1 4 0 · 3 9
─────────────
```

3.11 Subtracting decimal fractions

1 We can use place value partitioning to help us when we are subtracting decimal fractions. For example:

8·53 – 5·17

8 ones subtract 5 ones leaves **3 ones**

53 hundredths subtract 17 hundredths leaves **36 hundredths**

So, the answer is $3\frac{36}{100}$ = **3·36**

Use this method to calculate:

a) 5·74 – 3·16

b) 9·74 – 3·16

c) 28·56 – 8·43

d) 28·56 – 8·15

e) 52·93 – 12·08

f) 184·47 – 6·18

g) 285·82 – 9·73

2 We can use an empty number line and counting on when subtracting a decimal fraction from a whole number. For example:

13 − 5·18

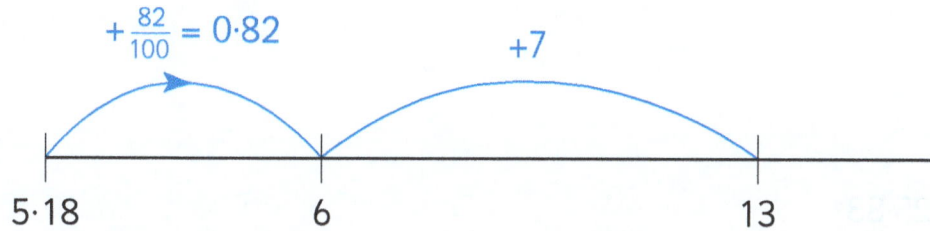

Count on from 5·18 to the next whole number, which is 6

5·18 + **0·82** = 6

Count on from 6 to 13, which is 7

6 + **7** = 13

So, 13 − 5·18 = 7·82

Use this method to calculate:

a) 12 − 4·82

b) 12 − 7·82

c) 12 − 7·39

d) 40 − 25·16

e) 43 − 25·16

f) 43 − 25·83

g) 43 − 35·83

h) 100 − 6·37

i) 500 − 45·81

j) 200 − 71·12

k) 700 − 94·54

l) 1000 − 38·62

3 a) Use the numbers on these starbursts to write six different subtractions.
For example, 2·73 − 1·06.

 **1·06**  **4**  **2·73**  **6·59**

b) Now calculate the answers to the subtractions you have made.
Explain the strategy you used for each one.

★ **Challenge**

Find three different pairs of numbers, each with a difference of 3·68.

3.12 Subtracting decimal fractions using standard algorithms

1 We can use a standard written algorithm to help us subtract decimal fractions that are too tricky to work out mentally. For example:

43·53 – 16·27

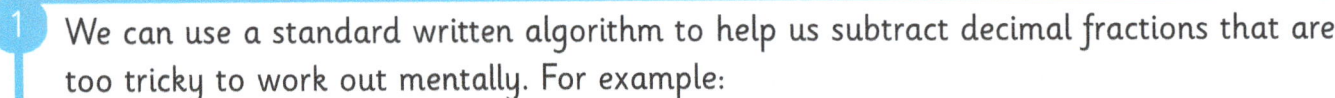

Use a standard written algorithm to calculate the following:

a) 36·53 – 17·61

b) 36·53 – 17·68

c) 36·19 – 17·68

d) 52·74 – 8·16

e) 52·04 – 8·16

f) 52·04 – 8·52

g) 73·82 – 6·92

h) 48·05 – 19·06

i) 29·17 – 23·86

j) 94·66 – 68·17

k) 50·05 – 33·82

l) 29·14 – 26·76

m) 38·04 – 28·77

n) 90·84 – 79·64

o) 63·91 – 39·96

2 Find the errors in this written algorithm and correct them.

$$
\begin{array}{r}
1\,7 \cdot 4\,6 \\
-\ 1\,1 \cdot 9\,1 \\
\hline
6 \cdot 5\,5 \\
\end{array}
$$

3 What is the missing number in this calculation?

17·27 – ☐ = 1·89

1. Fill in the missing digits in each of these subtraction calculations.

a)

```
    8 □ · 4 □
  -   □ 2 · □ 7
  ─────────────
    4 7 · 4 9
```

b)

```
    9 2 · □ □
  -   □ □ · 8 6
  ─────────────
    3 4 · 5 7
```

2. Now make up two mystery number questions and challenge a partner to solve them.

1 Work out the answers to each of these additions using the most efficient method. For each one, say which method you used.

a) $14 \cdot 73 + 3 \cdot 2$

b) $25 \cdot 46 + 9 \cdot 6$

c) $6 \cdot 4 + 34 \cdot 03$

d) $12 \cdot 7 + 26 \cdot 29$

e) $67 \cdot 4 + 35 \cdot 83$

f) $31 \cdot 98 + 4 \cdot 8$

g) $136 \cdot 2 + 5 \cdot 67$

h) $400 \cdot 06 + 0 \cdot 51$

i) $207 \cdot 42 + 30 \cdot 79$

j) $6 \cdot 83 + 5 \cdot 7 + 3 \cdot 41$

k) $4 \cdot 5 + 8 \cdot 8 + 10 \cdot 4 + 17 \cdot 52$

2 Work out the answers to each of these subtractions using the most efficient method. For each one, say which method you used.

a) 6·1 – 3·85

b) 7·6 – 6·82

c) 4·1 – 0·02

d) 34·48 – 23·2

e) 16·4 – 2·98

f) 70·7 – 32·01

g) 45·76 – 30·4

h) 41·5 – 19·83

i) 26·08 – 4·3

j) 300 – 60·7

k) 150·9 – 14·61

l) 604·26 – 48·4

3 Write in the arrow what you must add or subtract from each input number to reach the output number. The first one has been done for you.

	INPUT		OUTPUT		INPUT		OUTPUT

a) 65·34 → +1 → 66·34   b) 53·76 → → 53·56

c) 18·71 → → 18·77   d) 44·03 → → 44·2

e) 83·89 → → 83·69   f) 27·31 → → 28·61

g) 51·8 → → 50·62   h) 19·78 → → 23·89

i) 39·69 → → 33·62   j) 163·01 → → 161·51

★ **Challenge**

Using each of the digits 1, 2, 3, 4, 5, 6, 7, 8 and 9 only once, complete this calculation.

```
    0 . ☐ ☐ ☐
    0 . ☐ ☐ ☐
+   0 . ☐ ☐ ☐
  _____
    0 . 9 9 9
  _____
```

3.14 Representing word problems involving decimal fractions

Complete the Think Board for each problem and solve it.

1

Word problem	Bar model
Three packages in a delivery truck weigh 56·13kg, 7·08kg and 20·94kg. What is the total weight of all three packages?	

Answer

Empty number line	Calculation(s)

2

Word problem	Bar model
Kerry and Lauren are measuring how far they can throw a bean bag. Kerry's throw measures 3·63m and Lauren's throw measures 4·17m. How much further did Lauren throw the bean bag?	

Answer

Empty number line	Calculation(s)

3

Word problem

In an athletics competition the winner in the high jump set a new record by jumping 0·06m higher than the old record of 1·58m. What is the new high jump record?

Bar model

Answer

Empty number line

Calculation(s)

4

Word problem

Some students grew three sunflowers in the school greenhouse. When they measured them they noted that the total height of the three sunflowers was 5·81m. Two of the sunflowers measured 2·07m and 1·92m. What was the height of the third sunflower?

Bar model

Answer

Empty number line

Calculation(s)

Write a word problem for these partially completed Think Boards. Complete the Think Board for your word problem.

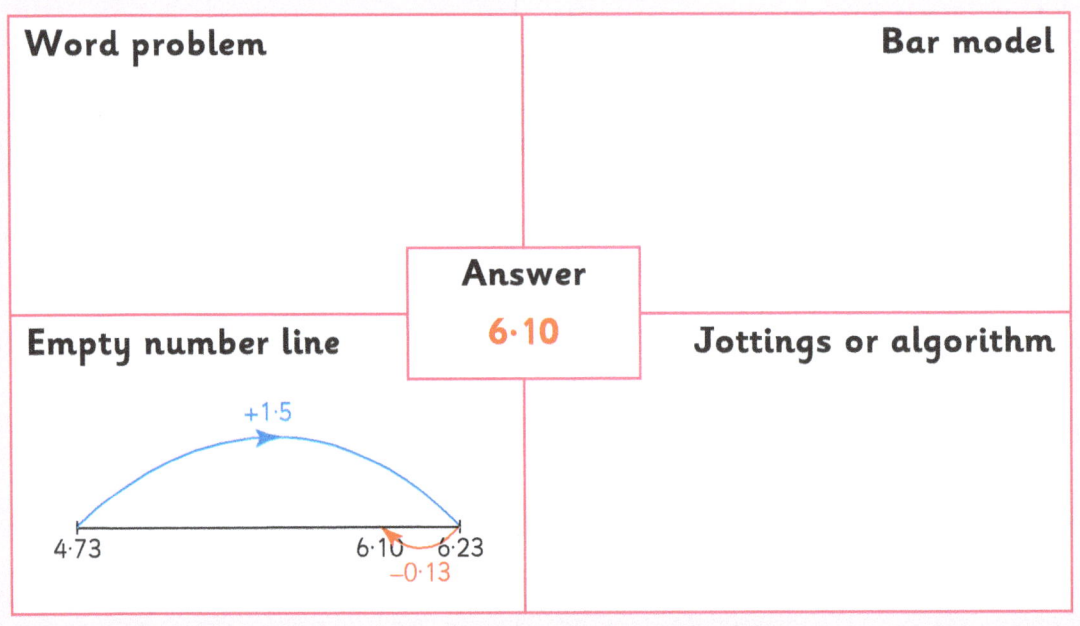

Word problem

Bar model

Answer
6·10

Empty number line

+1·5

4·73   6·10   6·23
        −0·13

Jottings or algorithm

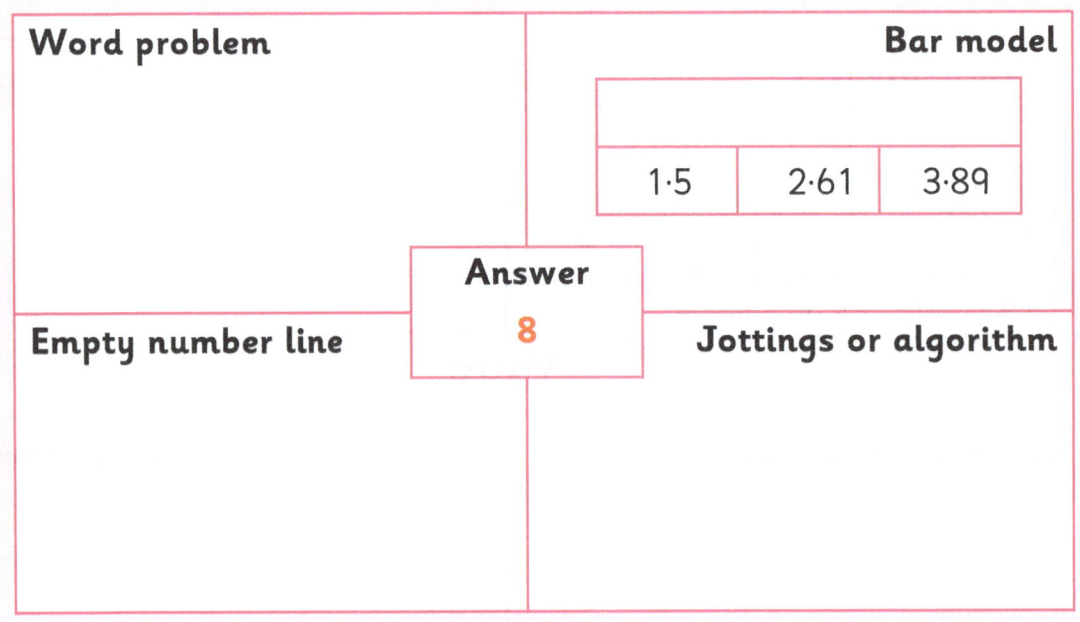

Word problem

Bar model

| 1·5 | 2·61 | 3·89 |

Answer
8

Empty number line

Jottings or algorithm

1 In a school athletics competition, the winning time in the 100m relay was 53·93 seconds. Each team had four runners. In the winning team, the first three runners' times were 13·46 seconds, 13·20 seconds and 14·01 seconds. What was the fourth runner's time?

2 325m of fencing has been ordered for a zoo. 45·8m of fencing is needed for the small animal area. The picnic area and the play equipment area each need 84·6m of fencing. The car park needs 118·5m of fencing around it. Has enough fencing been ordered? Explain your answer.

3 Pencils, pens and rulers are ordered to make up packs of equipment to sell at a school fair. Their order came to £203·28. The pencils cost £65·70 and the pens cost £11·95 more than the pencils. The rulers cost £17·72 less than the pens. How much was each item on the order?

4 Scientists at four Scottish weather stations measured the amount of rain that fell in July and got a combined total of 273·1mm. 71·55mm of rain was measured at Orkney and 49·28mm was measured at Dunbar. The scientists noticed that the amount of rain measured at Aviemore was double that of Dunbar. The fourth weather station was in Edinburgh. Work out how much rain was measured at the weather station in Edinburgh in July.

In this magic square, the numbers in each of the rows, columns and diagonals add up to 3·9. Fill in the blanks in the magic square.

0·66	0·6	0·24	1·38	
	0·72	0·36		1·44
	1·14	0·78		0·06
		1·2		0·48
	0·18		0·96	

4.1 Multiplication and division facts for 7

1 Use the multiplication or division fact that is given to help you work out the answer to the problem. Two have been done for you.

a) $2 \times 7 = 14$, so $4 \times 7 =$ 28

b) $6 \times 7 = 42$, so $3 \times 7 =$

c) $10 \times 7 = 70$, so $5 \times 7 =$

d) $4 \times 7 = 28$, so $8 \times 7 =$

e) $5 \times 7 = 35$, so 35 $\div 7 = 5$

f) $7 \times 7 = 49$, so ⬚ $\div 7 = 7$

g) $3 \times 7 = 21$, so ⬚ $\div 7 = 3$

h) $9 \times 7 = 63$, so ⬚ $\div 7 = 9$

2 Fill in the missing number in each of these.

a) $7 \times 4 =$

b) $7 \times$ ⬚ $= 28$

c) $14 \div 7 =$

d) $7 \times$ ⬚ $= 14$

e) $42 \div 7 =$

f) ⬚ $\div 7 = 10$

g) $56 \div 7 =$

h) ⬚ $\times 7 = 14$

i) 12 × 7 = ____

j) ____ × 7 = 0

k) ____ ÷ 7 = 6

l) ____ × 7 = 63

3 We can use one 7s multiplication fact to work out other number facts, like this:

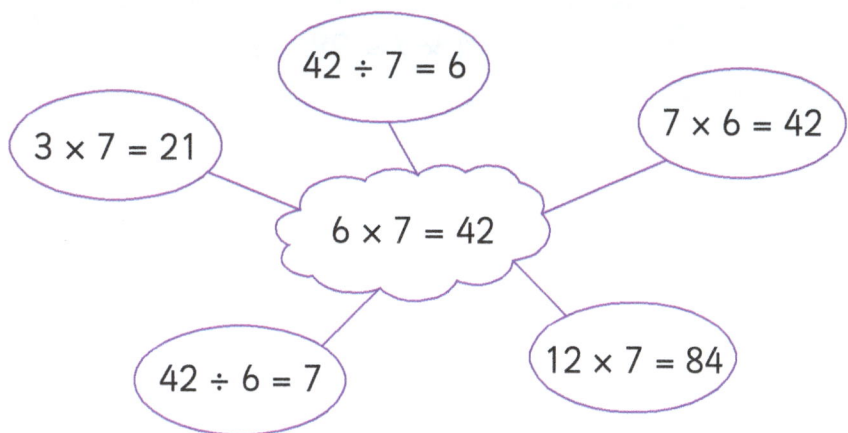

3 × 7 = 21   42 ÷ 7 = 6   7 × 6 = 42   6 × 7 = 42   42 ÷ 6 = 7   12 × 7 = 84

Add as many facts as you can to each of these.

a)

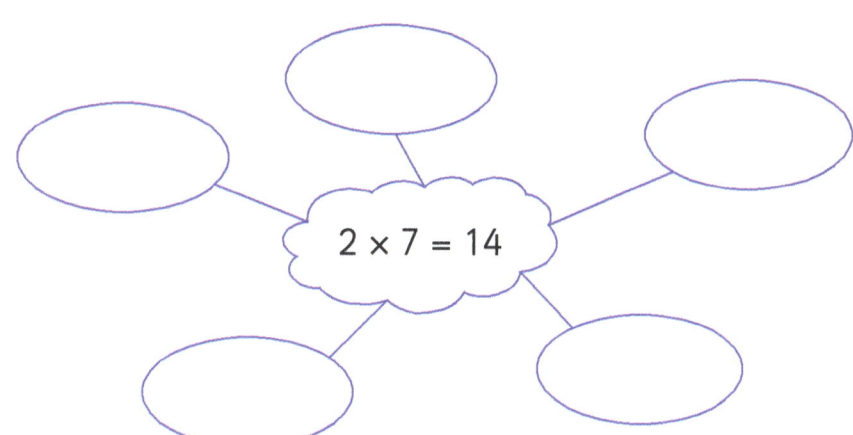

2 × 7 = 14

b)

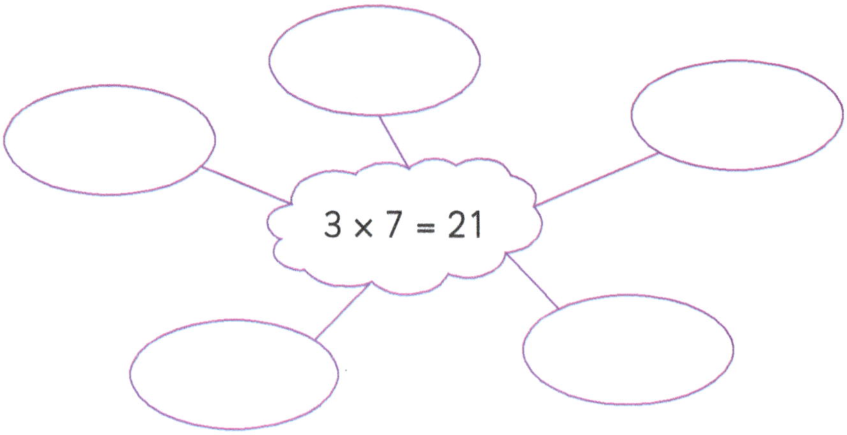

3 × 7 = 21

c)

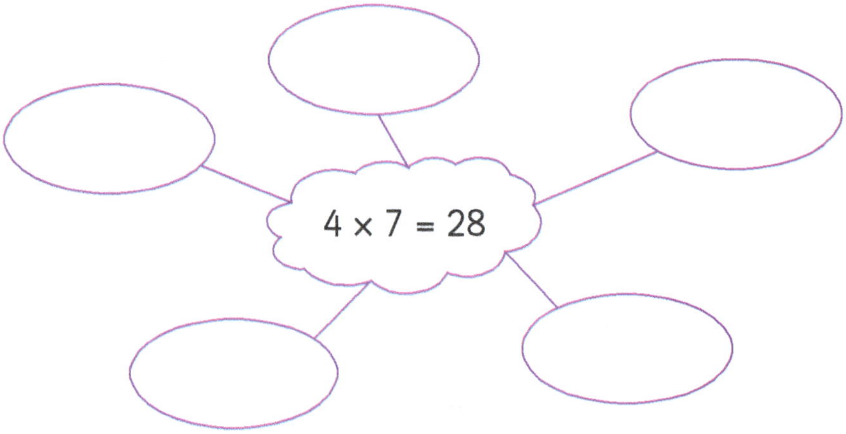

4 × 7 = 28

d)

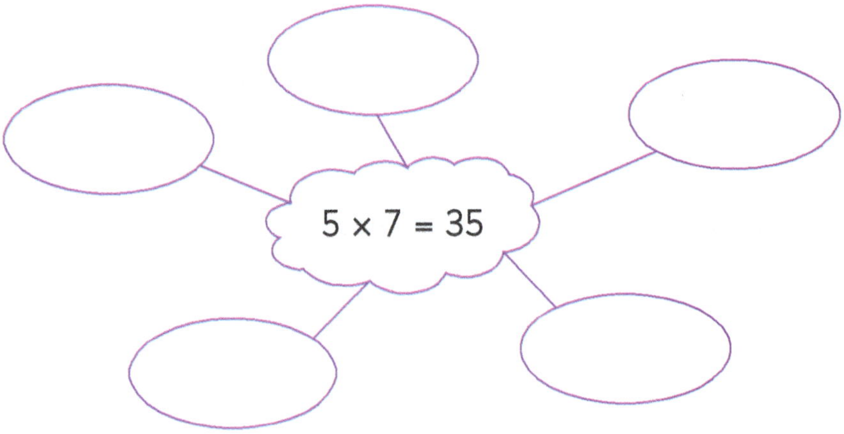

5 × 7 = 35

e)

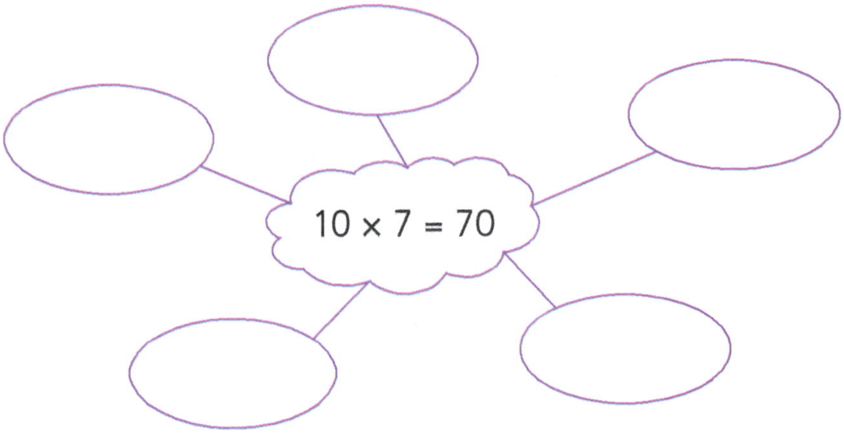

10 × 7 = 70

1. Complete this number puzzle using multiplication facts for 7. You can only enter one digit or one symbol into each square.

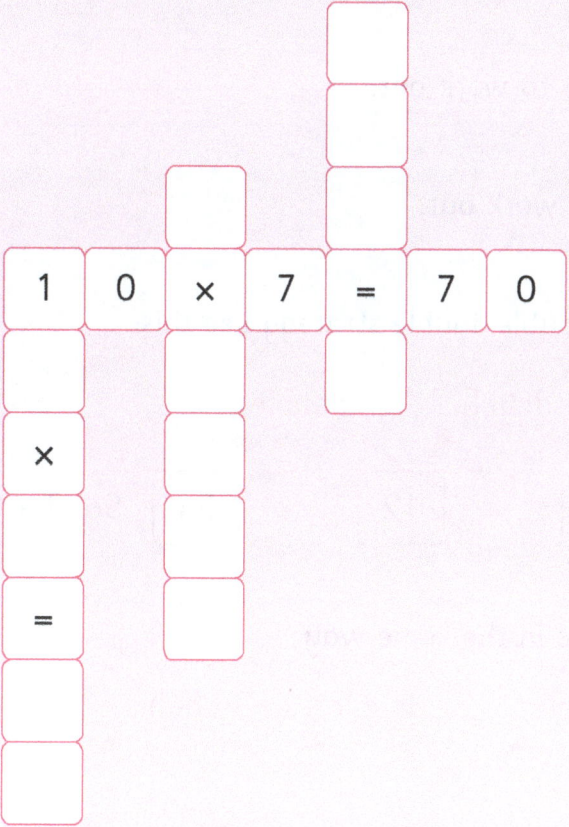

2. Can you find a different way to complete the puzzle?

4.2 Recalling multiplication and division facts for 8

We can use the double-double-double method to help us when we are multiplying by 8. For example, to find 3 × 8 we begin by doubling 3:

3 × 2 = **6**

Then we double the answer to work out:

3 × 4 = **12**

We then double it again to work out:

3 × 8 = **24**

We can show the double-double-double strategy like this:

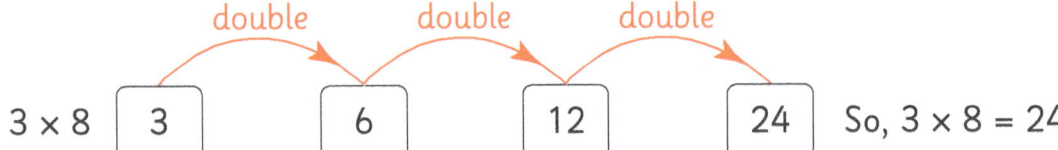

3 × 8 [3] —double→ [6] —double→ [12] —double→ [24] So, 3 × 8 = 24

1 Complete each of these in the same way.

a) 4 × 8 [4] —double→ [ ] —double→ [ ] —double→ [ ] So, 4 × 8 = ◯

b) 5 × 8 [5] —double→ [ ] —double→ [ ] —double→ [ ] So, 5 × 8 = ◯

c) 7 × 8 [7] —double→ [ ] —double→ [ ] —double→ [ ] So, 7 × 8 = ◯

d) 2 × 8 [2] —double→ [ ] —double→ [ ] —double→ [ ] So, 2 × 8 = ◯

e) 6 × 8 [6] —double→ [ ] —double→ [ ] —double→ [ ] So, 6 × 8 = ◯

2 Complete these to work out the multiplication facts for 8. The first one has been done for you.

a) (5) × 8

double — double — double

| 5 | 10 | 20 | 40 | 5 × 8 = 40 |

b) ( ) × 8

double — double — double

| | | 12 | | |

c) ( ) × 8

double — double — double

| 8 | | | | |

d) ( ) × 8

double — double — double

| | 24 | | | |

e) ( ) × 8

double — double — double

| | | 28 | | |

f) ( ) × 8

double — double — double

| | | 44 | | |

3 We can write two multiplication facts and two division facts using the numbers in this triangle.

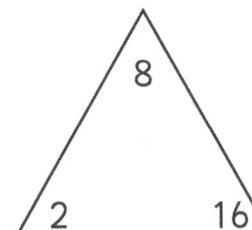

$2 \times 8 = 16$

$8 \times 2 = 16$

$16 \div 8 = 2$

$16 \div 2 = 8$

Use the numbers in these triangles to write two multiplication facts and two division facts.

a)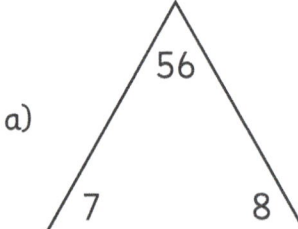

_____ × _____ = _____

_____ × _____ = _____

_____ ÷ _____ = _____

_____ ÷ _____ = _____

b)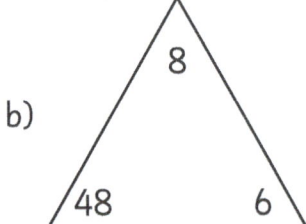

_____ × _____ = _____

_____ × _____ = _____

_____ ÷ _____ = _____

_____ ÷ _____ = _____

c)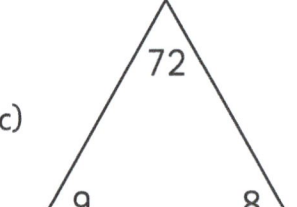

_____ × _____ = _____

_____ × _____ = _____

_____ ÷ _____ = _____

_____ ÷ _____ = _____

d)

_____ × _____ = _____

_____ × _____ = _____

_____ ÷ _____ = _____

_____ ÷ _____ = _____

Work out how this multiplication wheel has been completed.

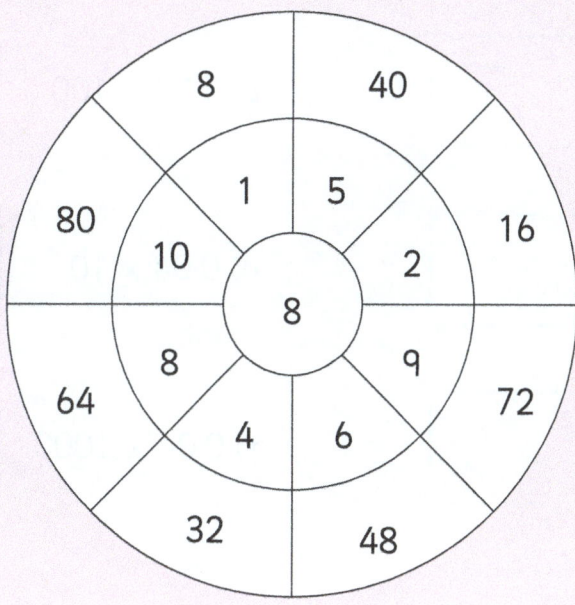

Now complete this multiplication wheel.

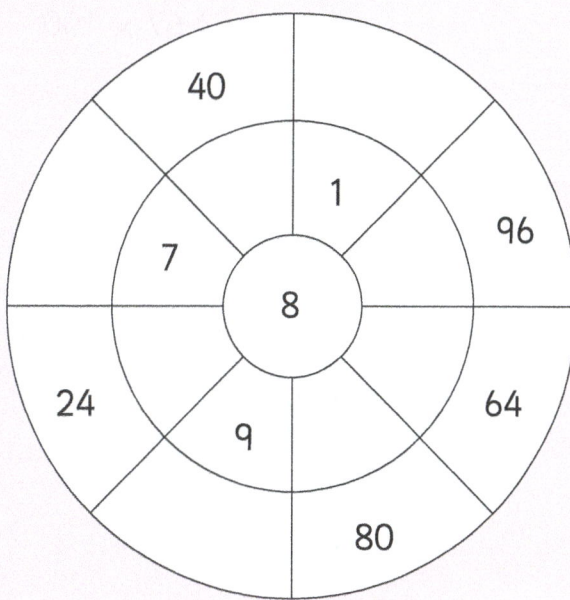

4.3 Multiplication with decimal fractions

1 Solve these problems:

a) 3·27 × 10

b) 3·27 × 100

c) 3·27 × 1000

d) 0·04 × 10

e) 0·04 × 100

f) 0·04 × 1000

g) 26·81 × 10

h) 6·49 × 100

i) 0·62 × 1000

j) 84·67 × 100

2 Circle True or False for each statement. Change one number in those that are false to make a true statement. Write this true statement in the answer box.

a) 6·18 × 100 = 618   **True**   **False**

b) 0·47 × 10 = 47   **True**   **False**

c) 83·25 × 1000 = 8325   **True**   **False**

d) 0·13 × 10 = 1.3   **True**   **False**

e) 9·03 × 1000 = 9030   **True**   **False**

f) 194·3 × 100 = 1943   **True**   **False**

g) 1000 × 4·31 = 4000.31   **True**   **False**

h) 10 × 3266·45 = 326 645   **True**   **False**

3   What number should be under the splodge?

a) 1·96 × ▨ = 196

b) 14·72 × ▨ = 147·2

c) 49·94 × ▨ = 49 940

d) ▨ × 8·33 = 833

e) ▨ × 90·07 = 900·7

f) 0·199 × ▨ = 19·9

★ **Challenge**

Match each question to its answer. One has been done for you.

Question	Answer
0·048 × 10	48
4·8 × 10	0·48
480 × 100	4·8
0·48 × 1000	4800
48 × 100	48 000
0·48 × 10	480

4.4 Division with decimal fractions

1 Solve these questions:

a) 83 ÷ 10

b) 83 ÷ 100

c) 83 ÷ 1000

d) 7 ÷ 10

e) 7 ÷ 100

f) 7 ÷ 1000

g) 30·71 ÷ 1000

h) 1·9 ÷ 100

2 Circle True or False for each statement. Change one number in those that are false to make a true statement. Write this true statement in the answer box.

a) 304 ÷ 100 = 3·4   **True**   **False**

b) 0.58 ÷ 10 = 0·058   **True**   **False**

c) 15 467 ÷ 1000 = 1·5467   **True**   **False**

d) 25·53 ÷ 10 = 255·3   **True**   **False**

e) 195·4 ÷ 100 = 1·954   **True**   **False**

f) 1·88 ÷ 100 = 0·188   **True**   **False**

3 Fill in the missing numbers in each of these:

a) $45 \div \boxed{} = 0.45$

b) $\boxed{} \div 10 = 13.7$

c) $3400 \div \boxed{} = 3.4$

d) $1.54 \div \boxed{} = 0.154$

e) $\boxed{} \div 100 = 164.32$

f) $0.54 \div \boxed{} = 0.054$

★ Challenge

1. Fill in the missing numbers in these number machines.

$\boxed{} \xrightarrow{\div 1000} \boxed{} \xrightarrow{\times 2} \boxed{} \xrightarrow{\div 10} \mathbf{0.54}$

$\boxed{} \xrightarrow{\div 10} \boxed{} \xrightarrow{\times 4} \boxed{} \xrightarrow{\div 100} \mathbf{1.68}$

2. Now create two number machines of your own. The answers must be numbers that have tenths and hundredths in them.

$\boxed{} \xrightarrow{\div 1000} \boxed{} \xrightarrow{\times 2} \boxed{} \xrightarrow{\div 10} \boxed{}$

$\boxed{} \xrightarrow{\div 10} \boxed{} \xrightarrow{\times 4} \boxed{} \xrightarrow{\div 100} \boxed{}$

4.5 Solving multiplication problems

1 We can use the grid method to partition numbers by place value and solve multiplication problems. For example:

3216 × 4

×	3000	200	10	6
4	12000	800	40	24

12000 + 800 + 40 + 24 = 12864

Answer each of these questions using the grid method.

a) 5283 × 3

×				

b) 5283 × 8

×				

c) 1294 × 7

×				

d) 3294 × 7

×				

e) 6231 × 6

×				

f) 4883 × 5

×				

2

We can use brackets to partition numbers by place value and solve multiplication problems. For example:

$3216 × 4 = (3000 × 4) + (200 × 4) + (10 × 4) + (6 × 4)$

$= 12\,000 + 800 + 40 + 24$

$= 12\,864$

Answer these questions using brackets.

a) 4825 × 3

b) 4825 × 7

c) 6133 × 4

d) 2133 × 4

e) 8056 × 6

f) 2671 × 9

3

a) A baker sells boxes of cupcakes. There are eight cupcakes in each box. If the baker sells 2476 boxes in one day, how many cupcakes are sold altogether?

b) The baker reduces the size of the cupcake boxes so that now there are only six cupcakes in each box. If the baker sells 3281 boxes in a day, how many cupcakes are sold altogether?

c) A large order of cupcakes is packed into seven crates and loaded onto a van. If each crate contains 4913 boxes, how many boxes are there in the order altogether?

⭐ **Challenge**

Use the digits 0 to 9 to complete the multiplication below.
Each digit can only be used once.

| 0 | 1 | 2 | 3 | 4 | 5 | 6 | 7 | 8 | 9 |

| | | | 4 | × | 3 | = | | 7 | 0 | | |

1 We can use the grid method to partition decimal fractions by place value to make multiplying easier. For example:

6 × 8·23

×	8	0·2	0·03
6	48	1·2	0·18

48 + 1·2 + 0·18 = 49·38

Use the grid method to work out the answers to these problems.

a) 3 × 7·46

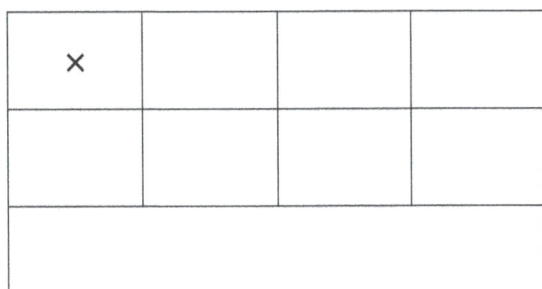

b) 8 × 7·46

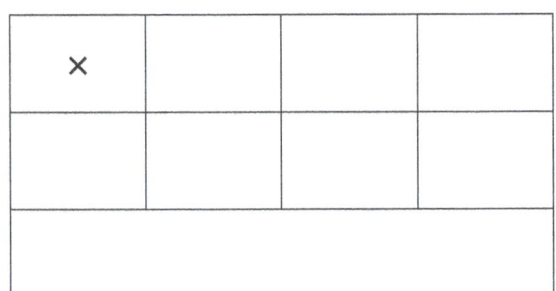

c) 8 × 4·32

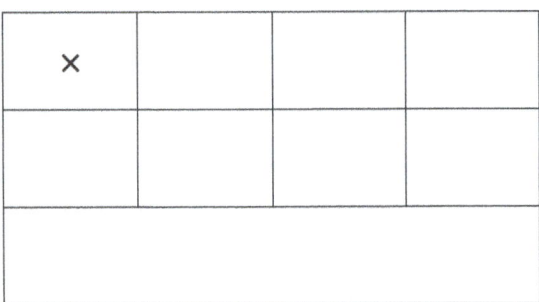

d) 5·44 × 6

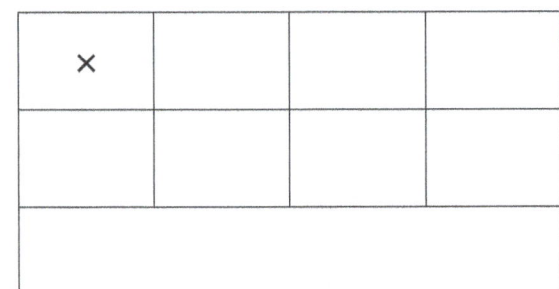

e) 1·87 × 4

f) 3.7 × 9

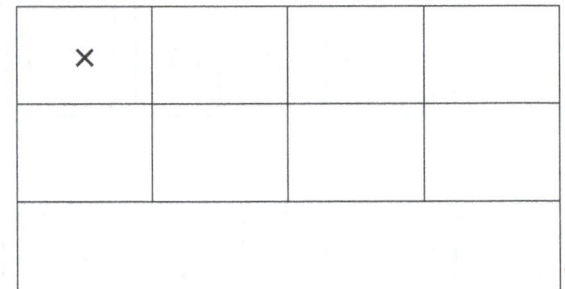

2 We can use brackets to partition decimal fractions by place value to make multiplying easier. For example:

$6 \times 8.23 = (6 \times 8) + (6 \times 0.2) + (6 \times 0.03)$
$= 48 + 1.2 + 0.18$
$= 48 + 1.38$
$= 49.38$

Use brackets to work out the answers to these problems.

a) 6×5.83

b) 6×3.83

c) 4×3.83

d) 6.28×7

3 Some students are ordering seeds to plant in the school garden. Work out the total cost for each of these collections.

a) Three packets of sunflower seeds at £2·34 per packet.

b) Eight packets of carrot seeds at £1·39 per packet.

c) Four packets of pumpkin seeds at £3·81 per packet.

d) Seven packets of radish seeds at £0·69 per packet.

e) Six packets of sweet pea seeds at £2·09 per packet.

Can you find the multiplier in each of these input/output machines?
One has been done for you.

Input		Output
3·91	× 4	15·64
8·46		76·14
56·7		283·5
16·9		135·2
0·83		5·81

4.7 Solving division problems using place value

1 We can use the grid method to partition numbers by place value and solve division problems. For example:

415 ÷ 5

÷	400	10	5
5	80	2	1

80 + 2 + 1 = 83

Use the grid method to work out the answers to these problems.

a) 284 ÷ 4

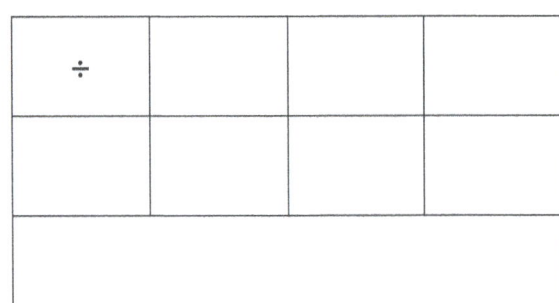

b) 468 ÷ 4

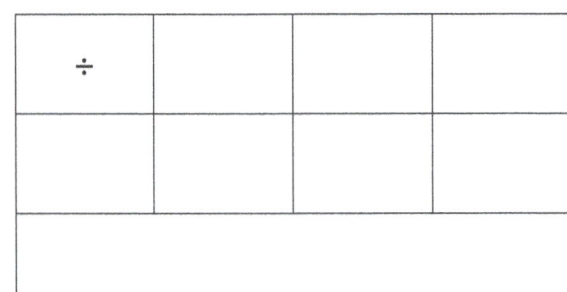

c) 366 ÷ 3

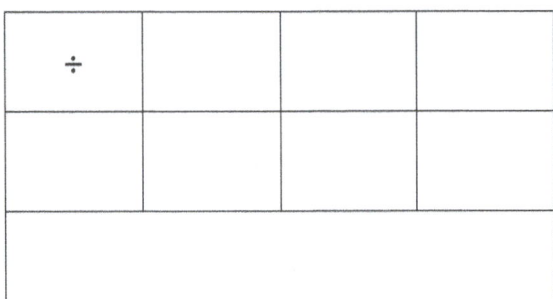

d) 633 ÷ 3

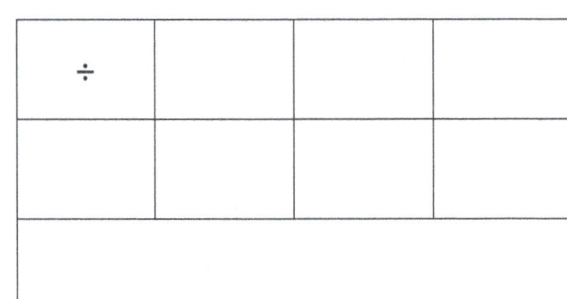

e) 945 ÷ 5

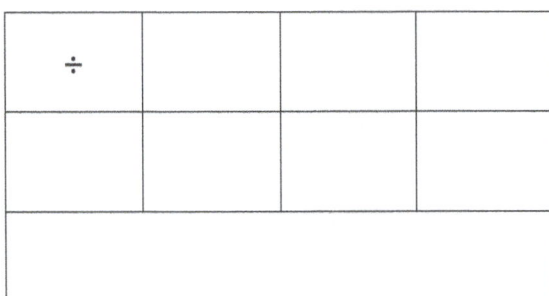

f) 396 ÷ 6

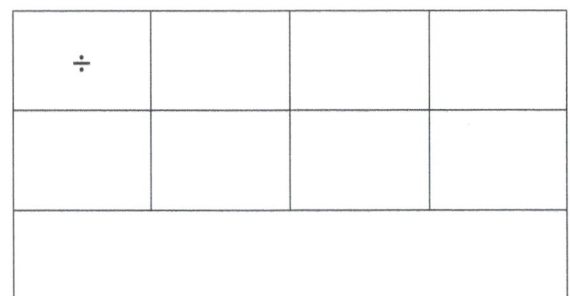

2 We can use multiplication facts that we know already to partition numbers and solve division problems. For example: $432 \div 6$

We know that 420 and 12 are both multiples of 6. We can then partition the 432 into 420 and 12 to make solving the problem easier.

$$432 \div 6 = (420 \div 6) + (12 \div 6)$$
$$= 70 + 2$$
$$= 72$$

Partition these numbers so that you can use multiplication facts. Then find the answers.

a) $365 \div 5$

b) $265 \div 5$

c) $256 \div 4$

d) $256 \div 8$

e) $642 \div 3$

f) $294 \div 7$

3 a) Choose your own strategy to divide these numbers by 9. Find the answers and say which strategy you used each time.

477	÷ 9 →		
288	÷ 9 →		
783	÷ 9 →		
612	÷ 9 →		

b) Choose your own strategy to divide these numbers by 8. Find the answer and say which strategy you used.

256	÷ 8 →		
472	÷ 8 →		
736	÷ 8 →		
504	÷ 8 →		

Some students are working together to find a number that can be divided exactly by 2, 3, 4, 5, 6, 7, 8 and 9. Cillian thinks 980 works. Niamh disagrees.

1. Do you agree with Niamh? Explain why.

2. Olivia thinks 2520 works. Is she correct? Show your thinking for the number 2520.

We can double one factor and halve the other in a multiplication problem and the answer will not change. For example:

5 × 14

double   halve

10 × 7   So 5 × 14 = 10 × 7 = 70

1 Find the answers to each of these using doubling and halving:

a) 3 × 18

b) 5 × 18

c) 4 × 22

d) 32 × 5

This also works for tripling and thirding. For example:

4 × 27

triple   third

12 × 9   So 4 × 27 = 12 × 9 = 108

2 Find the answers to each of these using tripling and thirding:

a) 3 × 18

b) 5 × 12

c) 6 × 15

d) 27 × 3

3 Use doubling and halving, or tripling and thirding, to help you solve these problems. Explain how you worked out each problem.

a) 16 sports tops were bought for school teams to use. Each top cost £7. How much did the tops cost altogether?

b) 21 tickets for a special screening at the cinema were bought for a youth club. Each ticket cost £4. How much did the tickets cost altogether?

c) Lunch boxes were packed into crates to take on a school outing. Each crate had 26 lunch boxes in it and there were five crates in total. How many lunch boxes were there altogether?

★ **Challenge**

How many pairs of equal calculations can you find here? Join these up by drawing a line between them. One has been done for you.

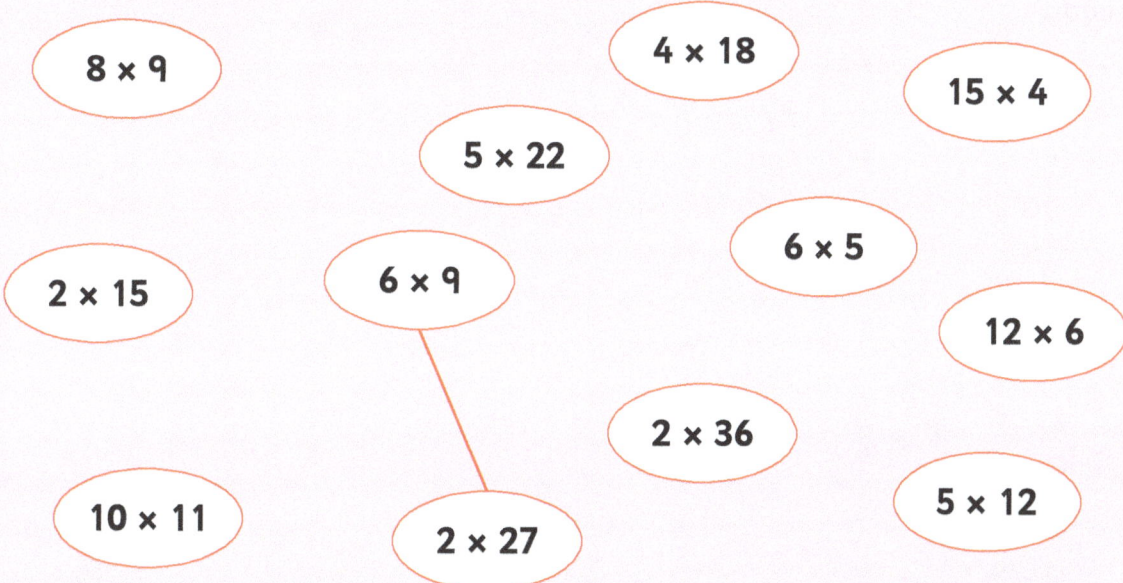

8 × 9

4 × 18

15 × 4

5 × 22

6 × 5

2 × 15

6 × 9

12 × 6

2 × 36

10 × 11

5 × 12

2 × 27

4.9 Solving division problems

1 Solve the following problems by rounding and compensating.
The first one has been done for you.

a) $56 \div 4 =$ | $(60 \div 4) - (4 \div 4)$ | $=$ | $15 - 1 = 14$

b) $81 \div 3 =$ | | $=$ |

c) $92 \div 4 =$ | | $=$ |

d) $57 \div 3 =$ | | $=$ |

e) $145 \div 5 =$ | | $=$ |

2 Eggs are being packed into boxes. Each box will have six eggs in it.
Use rounding and compensating to work out how many boxes will be needed
for 84 eggs.

3

a) Angus, Brodie, Robyn and Ava are working together on an art project. Their teacher gives them 68 sticks to use in the project and they decide to share them equally. Use rounding and compensating to work out how many sticks each of them will get.

b) Another student joins them so the group now has five students in it. The teacher gives them an extra 27 sticks so they now have 95 altogether.
Use rounding and compensating to work out how many sticks each student will get if they share them equally.

c) The five students are given two packs of wiggly eyes to use in their art project. One pack has 48 wiggly eyes in it and the other has 32. If they share all the wiggly eyes equally, how many will each student get?

Alexander thinks that the answer to each of these division problems is the same, but Craig disagrees. He says, "One of them is different."

Do you agree with Alexander or Craig? Explain your answer.

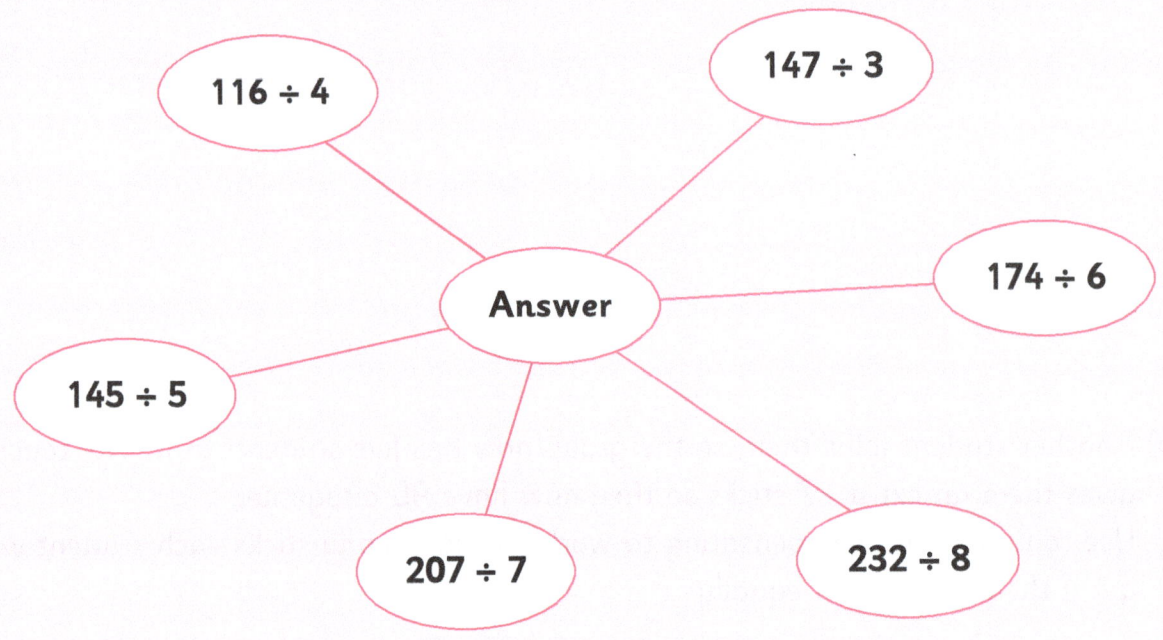

4.10 Solving multiplication problems

1 We can use the grid method to partition numbers to make multiplying easier. Complete these multiplication questions to find the answer.

a) 13 × 36

×	10	3
30		
6		

b) 23 × 36

×		

c) 23 × 46

×		

d) 23 × 44

×		

e) 64 × 28

×		

f) 53 × 37

×		

2 We can use a standard algorithm for multiplication to help us solve multiplication problems that are too challenging to solve mentally. For example:

23 × 36

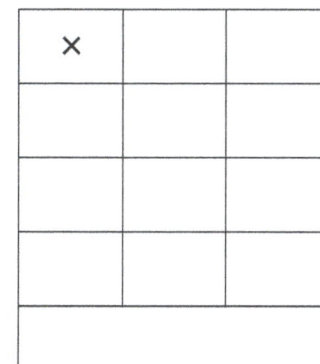

	36
	× 23
(36 × 3)	108
(36 × 20)	720
	828

Use the standard written method to solve these problems.

a) 32 × 43

b) 22 × 46

c) 41 × 29

3 Choose either the grid method or the standard written algorithm to solve each of these problems.

a) 14 boxes each contain 24 pencils. How many pencils are there altogether?

b) A theatre has 32 rows of seats with 26 seats in each row. How many seats are there altogether in the theatre?

Use the clues to complete this cross-number puzzle. Each box in the completed puzzle will only have one digit in it.

Across

2) 32 × 30 = [ ]

4) 11 × [ ] = 121

5) 10 × [ ] = 710

7) 13 × 7 = [ ]

9) [ ] × 12 = 780

10) 79 × 4 = [ ]

Down

1) 21 × 39 = [ ]

2) 7 × 13 = [ ]

3) 11 × [ ] = 737

6) [ ] × 2 = 316

8) 5 × [ ] = 55

9) 13 × [ ] = 858

4.11 Solving division problems using an algorithm

1 Use the written method to solve these division problems:

a) $4\overline{)\,9\;2}$

b) $4\overline{)\,7\;2}$

c) $3\overline{)\,7\;2}$

2 Work out the answer to these problems using the written form:

a) 95 ÷ 5

b) 135 ÷ 5

c) 185 ÷ 5

d) 168 ÷ 3

e) 152 ÷ 4

f) 203 ÷ 7

3 A craft group knitted 224 baby hats to donate to a local hospital. The hats are going to be packed into boxes before they are taken to the hospital. Use the written method to work out how many boxes will be needed if each box contains:

a) 4 hats

b) 7 hats

c) 8 hats

Some students are playing a game. They are trying to throw plastic balls into a large bowl. Each time they get a ball into the bowl they score 9 points. This table shows how many points each of them scored.

Carol	234
Sheona	153
Karen	207
Linda	261

Can you work out how many balls each of them got into the bowl?

4.12 Solving division problems involving decimal fractions

1 Use the written method to work out the answers to these problems.

a) $4\overline{)7 \cdot 28}$

b) $3\overline{)5 \cdot 28}$

c) $4\overline{)6 \cdot 12}$

2 Work out the answer to these problems using the written form:

a) 1·38 ÷ 3

b) 2·67 ÷ 3

c) 3·48 ÷ 6

d) 4·62 ÷ 6

e) 1·33 ÷ 7

f) 4·96 ÷ 8

3 Some students and their teacher are going to make soup. They go shopping for vegetables and this is their receipt.

CASH RECEIPT
Any Day Supermarket

Address:	5 High Street
Tel:	124 905 7385
Manager:	Noah Kane

4 large turnips	£4.92
9 carrots	£1.17
6 large potatoes	£7.38
2 leeks	£1.14
4 onions	£1.52
7 parsnips	£4.83

Price	£20.96
Sale	£3.96
Tax	£5.12

| Total | £30.24 |

Thank you for shopping!!!

VEGETABLES

Work out the cost of:

a) one large turnip

b) one carrot

c) one large potato

d) one leek

e) one onion

f) one parsnip

★ Challenge

Work out the missing digits in these calculations.

$$1 \cdot 5\,\boxed{}$$
$$4\,)\,\boxed{} \cdot 1\ 2$$

$$1 \cdot 7\,\boxed{}$$
$$2\,)\,\boxed{} \cdot 2\ 5$$

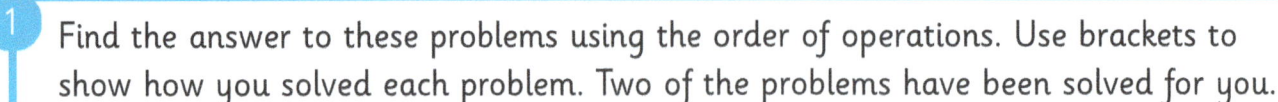

1 Find the answer to these problems using the order of operations. Use brackets to show how you solved each problem. Two of the problems have been solved for you.

a) 6 × 4 + 5

   = (6 × 4) + 5

   = 24 + 5

   = 29

b) 6 + 4 × 5

c) 6 + 4 × 5 − 3

d) 6 ÷ 3 + 12

   = (6 ÷ 3) + 12

   = 2 + 12

   = 14

e) 8 + 6 ÷ 3 − 1

f) 35 ÷ 7 + 20 − 6

g) 18 − 5 × 3 + 26

h) 29 + 4 × 6 − 15 ÷ 5

i) 28 ÷ 4 + 3 × 9 − 5

2 The answer given for each of these problems is incorrect. Use brackets and the order of operations to help you to find the errors, then work out the correct answer to each problem.

a) 18 + 9 ÷ 3 = 27 ÷ 3 = 9

b) 4 x 5 + 2 x 3 = 22 x 3 = 66

c) 37 − 7 x 3 + 8 = 37 − 21 + 8 = 8

d) 60 − 15 ÷ 3 + 2 x 4 = 45 ÷ 3 + 8 = 22

3 Write +, −, x or ÷ in each blank box so that each number sentence balances.

a) 4 ☐ 4 ☐ 3 = 16

b) 4 ☐ 6 ☐ 4 = 20

c) 4 ☐ 3 ☐ 3 = 13

d) 5 ☐ 4 ☐ 4 = 21

★ **Challenge**

1. Using each of the digits 4, 6, 7 and 8 once in this problem, what is the greatest answer you can find?

☐ + ☐ × ☐ − ☐ = ☐

2. Using each of the digits 4, 6, 7 and 8 once in this problem, what is the smallest answer you can find?

☐ + ☐ × ☐ − ☐ = ☐

5.1 Using knowledge of multiples and factors to work out divisibility rules

We know these divisibility rules:

A number will be divisible by

2	IF	the last digit is an even number, including zero
3	IF	the sum of the digits is a multiple of three
4	IF	the sum of the last two digits can be divided by four
5	IF	the last digit is zero or five
6	IF	the number is divisible by two **and** by three
8	IF	the sum of the last three digits is divisible by eight
9	IF	the sum of the digits is a multiple of nine
10	IF	the last digit is a zero

1 Write each of these numbers into the correct column in the table. Two have been done for you.

2312   3565   4158   4768   6246

6744   7808   8563   10128   14566

Divisible by 8	Not divisible by 8
2312	3565

2

a) Are these numbers divisible by 6? Circle all the numbers that can be divided exactly by 6.

| 254 | 336 | 521 | 582 | 690 | 879 |

| 1482 | 2581 | 3564 | 5892 | 21 246 | 28 136 |

3

Use the divisibility rules to work out if these statements are true or false. Circle **True** or **False** for each one and say why you have circled this. The first one has been done for you.

a) 684 is divisible by 3   (**True**)   **False**

6 + 8 + 4 = 18. This is a multiple of 3 so 684 is divisible by 3.

b) 684 is divisible by 9   **True**   **False**

c) 4575 is divisible by 4   **True**   **False**

d) 4575 is divisible by 5   **True**   **False**

e) 12 670 is divisible by 10   **True**   **False**

f) 12 670 is divisible by 8   **True**   **False**

g) 3414 is divisible by 6   **True**   **False**

h) 8577 is divisible by 9   **True**   **False**

★ Challenge

1. Use these clues to find the mystery number:
   - ❖ It is a three-digit number
   - ❖ It is divisible by 7
   - ❖ It is not divisible by 2
   - ❖ The sum of its digits is 4

2. Use these clues to find the mystery number:
   - ❖ It is a three-digit number less than 300
   - ❖ It is divisible by 2 and 5
   - ❖ It is not divisible by 3
   - ❖ The sum of its digits is 7

5.2 Using knowledge of multiples and factors to solve problems

1 A tennis club has 285 tennis balls that need to be put into tins for storage.

a) Each tin holds four tennis balls. Will the tennis balls all fit into the tins with none left over? How do you know?

b) How many tins will be needed altogether for the tennis balls?

c) If instead the club buys tins that hold 6 tennis balls, how many tins would be needed for the tennis balls?

2 The answers given for each of these calculations is incorrect. Use your knowledge of factors and multiples to explain why the answers cannot be correct. The first one has been done for you.

Calculation	Reason why the answer is incorrect
68 × 5 = 304	Multiples of 5 always end in 0 or a 5, not 4.
46 × 3 = 448	
280 × 9 = 2510	
265 ÷ 2 = 130·5	
424 ÷ 6 = 69	

3 A gardener has 48 paving slabs and is going to lay them in a rectangle in a new garden. The rectangle can be any length and width. Use your knowledge of factors to work out all the possible ways the gardener can lay all 48 paving slabs in the garden.

⭐ **Challenge**

Find the missing factors to complete these multiplication squares.

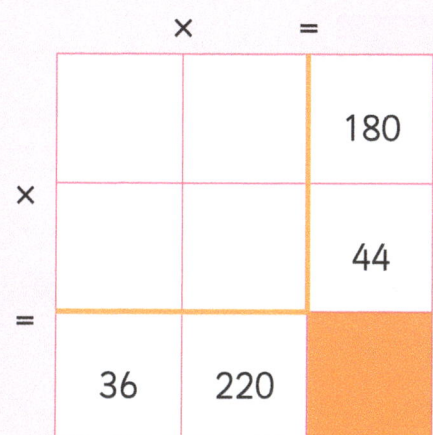

	×		=
			60
×	12		36
=	48	45	

	×		=
			180
×			44
=	36	220	

1 Convert each improper fraction into a mixed number. The first one has been done for you.

$$\frac{5}{2} = 2\frac{1}{2}$$

a)

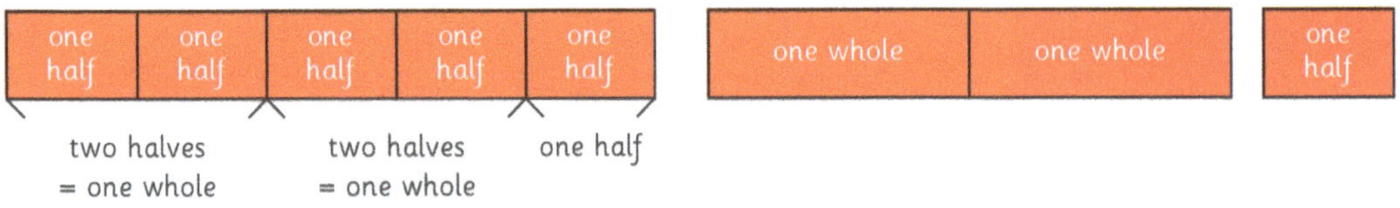

one half	one half	one half	one half	one half

two halves = one whole   two halves = one whole   one half

one whole	one whole

one half

$$\frac{5}{2} = 2\frac{1}{2}$$

b)

one third	one third	one third	one third	one third

$$\frac{\square}{\square} = \square\frac{\square}{\square}$$

c)

one quarter	one quarter	one quarter	one quarter	one quarter	one quarter	one quarter

$$\frac{\square}{\square} = \square\frac{\square}{\square}$$

d)

one fifth	one fifth	one fifth	one fifth	one fifth	one fifth	one fifth	one fifth	one fifth	one fifth	one fifth

$$\frac{\square}{\square} = \square\frac{\square}{\square}$$

2 Convert each mixed number into an improper fraction. The first one has been done for you.

$$2\frac{2}{3} = \frac{8}{3}$$

a)

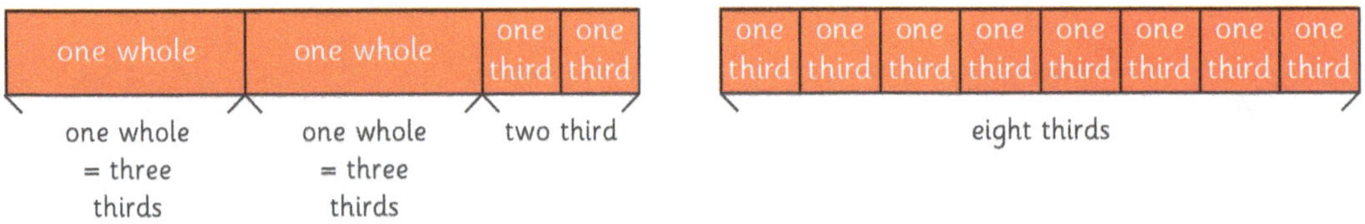

one whole	one whole	one third	one third

one whole = three thirds   one whole = three thirds   two third

one third	one third	one third	one third	one third	one third	one third	one third

eight thirds

$$2\frac{2}{3} = \frac{8}{3}$$

b)

one whole	one whole	one whole	one half

$$\square\frac{\square}{\square} = \frac{\square}{\square}$$

c)

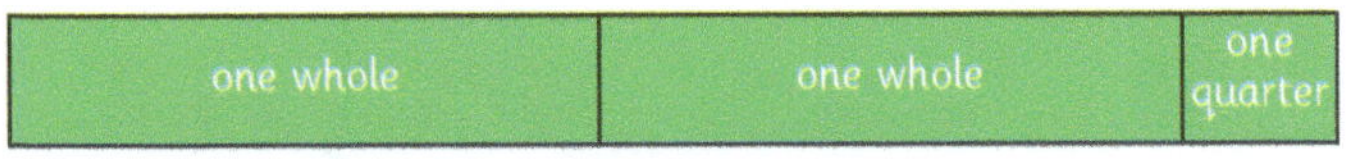

one whole	one whole	one quarter

$$\square\frac{\square}{\square} = \frac{\square}{\square}$$

d)

one whole	one fifth	one fifth	one fifth

$$\square\frac{\square}{\square} = \frac{\square}{\square}$$

3 Write the improper fraction and the mixed number for each shaded diagram. The first one has been done for you.

	Improper fraction	Mixed number
a)	$\dfrac{7}{4}$	$1\dfrac{3}{4}$
b)		
c)		
d)		

6.2 Comparing and ordering fractions

1. Write a common equivalent to help you solve the following:

   a) Sophie has completed three-quarters of a nature walk in a local park. Jack has completed four-fifths of the same walk. Who has walked further?

   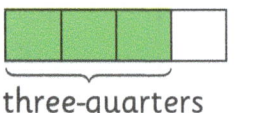 or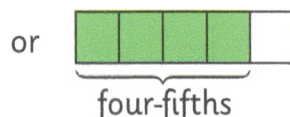

   three-quarters   four-fifths

   b) Lewis has filled three-quarters of a jar with sand. Isla has filled five-sixths of an identical jar with sand. Who has more sand in their jar?

   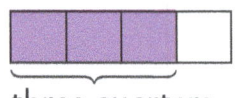 or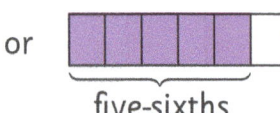

   three-quarters   five-sixths

2. Use equivalence to compare each pair of fractions. The first one has been done for you.

   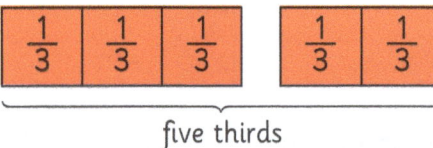

   five thirds

   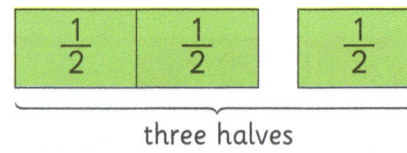

   three halves

   a) $\dfrac{5}{3}$ and $\dfrac{3}{2}$

   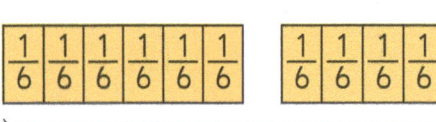

   ten sixths

   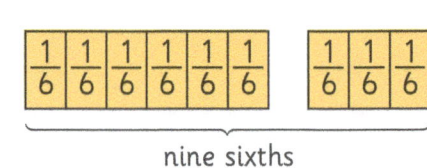

   nine sixths

   $\dfrac{10}{6}$ is greater than $\dfrac{9}{6}$ so $\dfrac{5}{3}$ is greater than $\dfrac{3}{2}$.

   b) $\dfrac{5}{2}$ and $\dfrac{7}{4}$

   c) $\dfrac{11}{5}$ and $\dfrac{5}{2}$

3 Write these improper fractions in the correct boxes on the number line:
$$\frac{15}{4}, \frac{16}{5}, \frac{7}{2}.$$

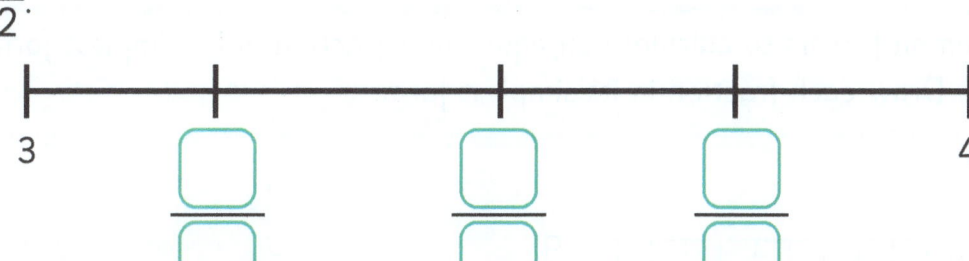

3   4

★ Challenge

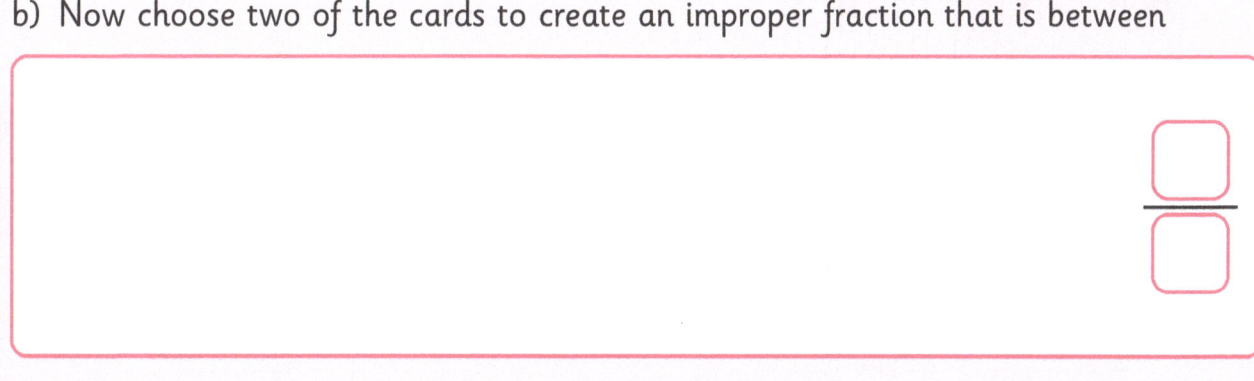

| 22 | 5 | 3 | 16 | 7 | 19 |

a) Choose two of the orange cards to create an improper fraction that is less than 2.

b) Now choose two of the cards to create an improper fraction that is between

c) Use two of the cards to create an improper fraction that is greater than 5.

6.3 Simplifying fractions

1 Use common factors to calculate an equivalent fraction in its simplest form for each of these. Draw each fraction in its simplest form.

a)

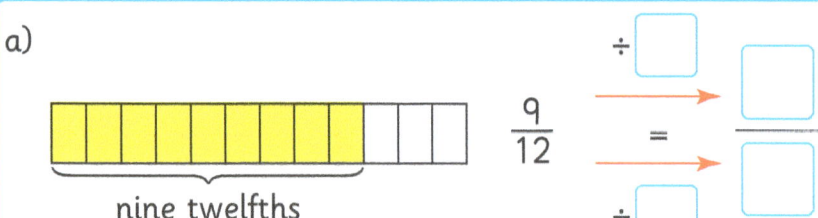

nine twelfths

$\frac{9}{12}$ $\div \square$
$=$
$\div \square$
$\frac{\square}{\square}$

b)

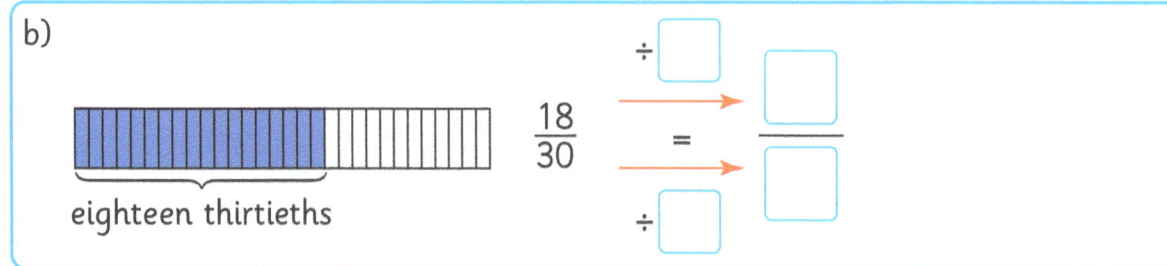

eighteen thirtieths

$\frac{18}{30}$ $\div \square$
$=$
$\div \square$
$\frac{\square}{\square}$

c)

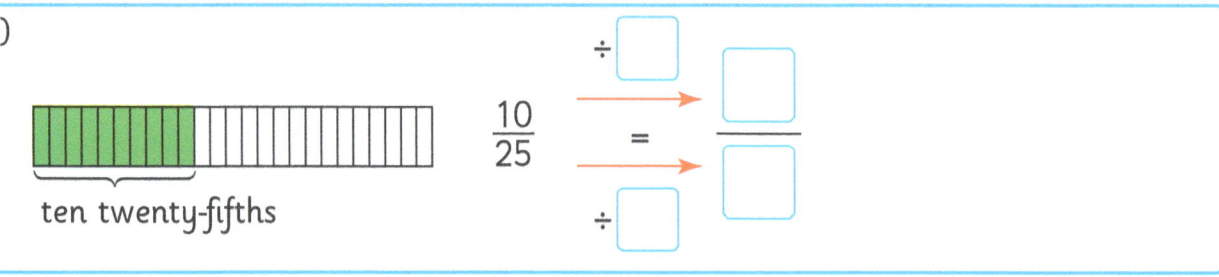

ten twenty-fifths

$\frac{10}{25}$ $\div \square$
$=$
$\div \square$
$\frac{\square}{\square}$

d)

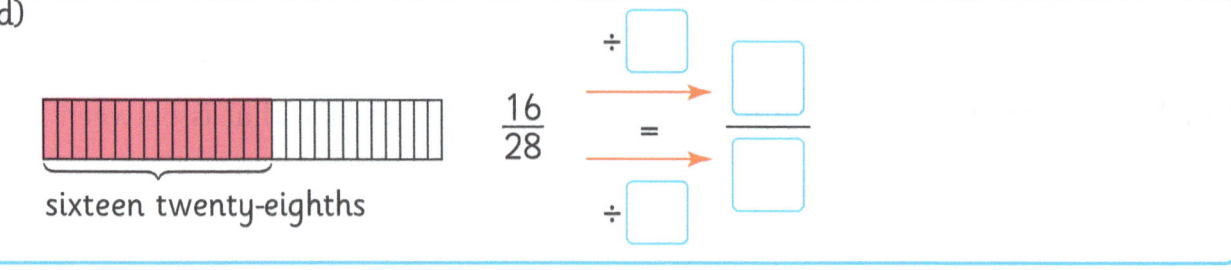

sixteen twenty-eighths

$\frac{16}{28}$ $\div \square$
$=$
$\div \square$
$\frac{\square}{\square}$

2 Use common factors to simplify each fraction. Remember to check if each answer is in its simplest form. The first one has been done for you.

a) $\frac{24}{40} \xrightarrow[\div 4]{\div 4} \frac{6}{10} \xrightarrow[\div 2]{\div 2} \frac{3}{5}$

b) $\frac{10}{40}$

c) $\frac{18}{27}$

d) $\frac{28}{36}$

e) $\dfrac{42}{60}$ 　　　　　　　　f) $\dfrac{32}{72}$

3 a) Circle the fractions that would be $\dfrac{3}{5}$ in their simplest form:

$\dfrac{12}{20}$　　$\dfrac{10}{15}$　　$\dfrac{30}{50}$　　$\dfrac{27}{45}$　　$\dfrac{21}{30}$

b) Circle the fractions that give $\dfrac{4}{7}$ when they are written in their simplest form:

$\dfrac{80}{140}$　　$\dfrac{32}{63}$　　$\dfrac{20}{35}$　　$\dfrac{24}{42}$　　$\dfrac{8}{14}$

★ Challenge

Circle the fraction that is the odd one out here. Explain your thinking.

$\dfrac{50}{80}$　　　　　$\dfrac{45}{72}$　　　　　$\dfrac{10}{16}$　　　　　$\dfrac{30}{42}$　　　　　$\dfrac{25}{40}$

6.4 Adding and subtracting fractions

1 We can use equivalence to help us to add and subtract fractions.

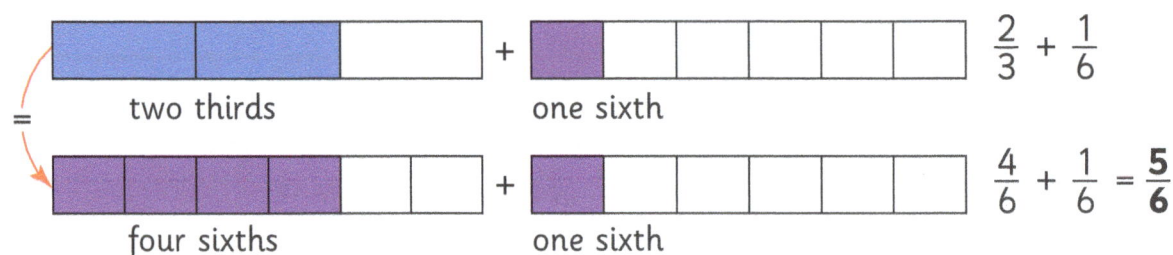

$$\frac{2}{3} + \frac{1}{6}$$

two thirds   one sixth

=

four sixths   one sixth

$$\frac{4}{6} + \frac{1}{6} = \frac{5}{6}$$

Use equivalence to solve the following:

a) one quarter   three eighths

$$\frac{1}{4} + \frac{3}{8} = \boxed{}$$

b) one quarter   five eighths

$$\frac{1}{4} + \frac{5}{8} = \boxed{}$$

c) two fifths   three tenths

$$\frac{2}{5} + \frac{3}{10} = \boxed{}$$

d) four thirds   five sixths

$$\frac{4}{3} - \frac{5}{6} = \boxed{}$$

2 Draw bar models to help solve the following:

a) $\frac{1}{2} + \frac{3}{8} = \boxed{}$

b) $\frac{11}{12} - \frac{2}{3} = \boxed{}$

c) $\frac{7}{10} + \frac{3}{5} = \boxed{}$

3 Draw diagrams to solve the following problems:

a) Fiona and Scott went fishing. Fiona caught a fish weighing $1\frac{2}{5}$ kg and Scott caught a fish weighing $1\frac{3}{10}$ kg. What is the total weight of the two fish?

b) A bottle contains $2\frac{5}{8}$ litres of water. Logan pours $1\frac{1}{4}$ litres of the water into a jug. How much water is left in the bottle?

c) Chloe is cycling around a track that is $4\frac{5}{6}$ km long. So far, she has cycled $2\frac{2}{3}$ km. How far does she still have to cycle?

★ Challenge

Can you place the fractions in the correct place on the grid to make each calculation true? You may only use each fraction once.

$\frac{3}{5}$ $\frac{11}{12}$ $\frac{8}{9}$

$\frac{7}{10}$ $\frac{1}{4}$ $\frac{1}{10}$

$\frac{2}{9}$

$\frac{5}{6}$ $\frac{3}{8}$

		+		=	$\frac{7}{10}$
$\frac{1}{8}$	+		=	
		−	$\frac{1}{5}$	=	$\frac{1}{2}$
		−		=	$\frac{1}{12}$
		+	$\frac{2}{3}$	=	

6.5 Converting decimal fractions to fractions

1 Draw lines to match the fraction shown by each bar to its decimal equivalent. One has been done for you.

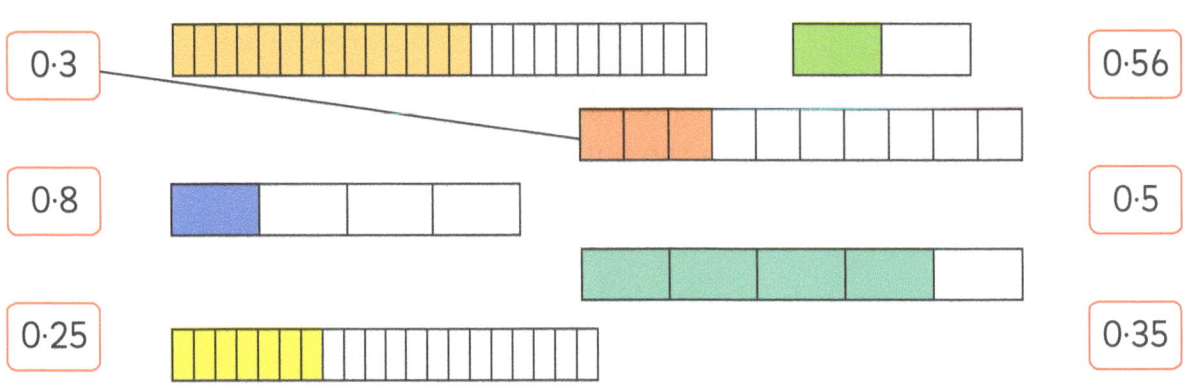

2 Write each decimal fraction as a fraction, then as a fraction in its simplest form. The first one has been done for you.

Decimal fraction	Fraction	Simplified fraction
0.75	$\frac{75}{100}$	$\frac{3}{4}$
0.45		
0.62		
0.18		
0.56		

3 Match each decimal fraction to an equivalent fraction by drawing lines.

Decimal fraction	0·65	0·7	0·75	0·66	0·6
Fraction	$\frac{3}{5}$	$\frac{7}{10}$	$\frac{33}{50}$	$\frac{3}{4}$	$\frac{13}{20}$

★ Challenge

a) Arrange these numbers in order from smallest to largest.

0·79 $\frac{4}{5}$ $\frac{67}{100}$ 0·7 $\frac{3}{4}$

Smallest ⟶ Largest

☐ ☐ ☐ ☐ ☐

Working

b) Now write a fraction on this card to add to the ordered list in part a. Your fraction must fit into the second place in the ordered list.

6.6 Calculating a fraction of a fraction

1 Draw a bar model to solve the following:

a) What is half of one-fifth of a bar?

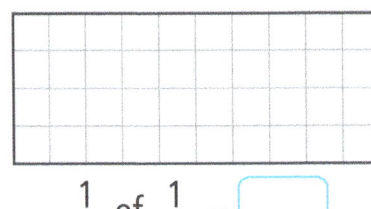

$$\frac{1}{2} \text{ of } \frac{1}{5} = \boxed{}$$

b) What is half of three-fifths of a bar?

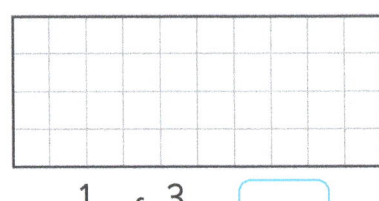

$$\frac{1}{2} \text{ of } \frac{3}{5} = \boxed{}$$

c) What is one quarter of three-fifths of a bar?

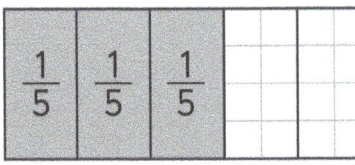
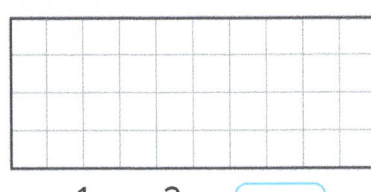

$$\frac{1}{4} \text{ of } \frac{3}{5} = \boxed{}$$

d) What is one third of three-quarters of a bar?

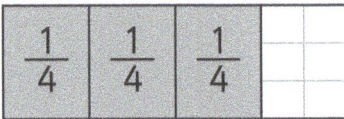
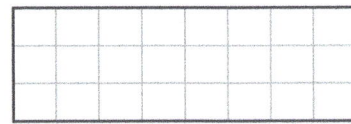

$$\frac{1}{3} \text{ of } \frac{3}{4} = \boxed{}$$

2 Draw a bar model to help you solve each of the following problems:

a) Connor has half a tub of paint. He uses one third of the paint during an art lesson. What fraction of a whole tub of paint did Connor use?

b) Romy and Zadie have one third of a bag of compost. They use three-quarters of the compost when planting some flowers. What fraction of a whole bag of compost did Romy and Zadie use?

c) Oli has two thirds of a roll of sticky tape. He uses one-quarter of the tape to wrap a parcel. What fraction of a whole roll of tape does Oli use?

★ **Challenge**

Olive says: "I think $\frac{3}{4}$ of $\frac{2}{5}$ is the same as $\frac{2}{5}$ of $\frac{3}{4}$."

Alice says: "No, that is not right. $\frac{3}{4}$ of $\frac{2}{5}$ is not the same as $\frac{2}{5}$ of $\frac{3}{4}$."

Do you agree with Olive or Alice? Use bar models to help you explain your answer.

6.7 Dividing a fraction by a whole number

1 Use the blank bar models to work out how the following could be shared out equally. The first one has been done for you.

a) $\frac{1}{3}$ Shared between two people $\frac{1}{6}$ $\frac{1}{6}$ $\frac{1}{3} \div 2 = \frac{1}{6}$ each

b) $\frac{1}{4}$ Shared between two people   _____

c) $\frac{1}{5}$ Shared between two people   _____

d) $\frac{1}{5}$ Shared between three people   _____

2 Use the blank bar models to work out how the following could be shared out equally.

a) $\frac{1}{4}$ $\frac{1}{4}$ $\frac{1}{4}$ Shared between two people   _____

b) $\frac{1}{3}$ $\frac{1}{3}$ Shared between two people   _____

c) $\frac{1}{5}$ $\frac{1}{5}$ $\frac{1}{5}$ Shared between three people   _____

d) $\frac{1}{6}$ $\frac{1}{6}$ $\frac{1}{6}$ $\frac{1}{6}$ $\frac{1}{6}$ Shared between four people   _____

3 Harry, Ethan and Brooke are sharing $\frac{5}{8}$ of a cake equally. What fraction of the cake do they each get?

★ **Challenge**

Work out the missing digits in these calculations.

a) $\dfrac{3}{\boxed{}} \div 4 = \dfrac{3}{20}$

b) $\dfrac{\boxed{}}{7} \div 2 = \dfrac{2}{7}$

c) $\dfrac{\boxed{}}{\boxed{}} \div 5 = \dfrac{3}{20}$

d) Make up two questions like these with missing digits.

Your questions should involve dividing a proper fraction by a whole number.

6.8 Dividing a whole number by a fraction

1. The food technology class are preparing fruit to make fruit pots for lunch. Calculate how many portions they can make from the following:

a) Five bananas to be divided into $\frac{1}{2}$ portions.

b) Seven apricots to be divided into $\frac{1}{2}$ portions.

c) Six mangos to be divided into $\frac{1}{2}$ portions.

d) Six pineapples to be divided into $\frac{1}{3}$ portions.

e) Four grapefruits to be divided into $\frac{1}{3}$ portions.

f) Eight pears to be divided into $\frac{1}{4}$ portions.

g) Nine melons to be divided into $\frac{1}{5}$ portions.

2

a) A chef has 5 kilograms of pizza dough. If one pizza needs $\frac{1}{4}$ of a kilogram of pizza dough, how many pizzas can he make?

b) The chef has 12 metres of foil to use for wrapping pizzas. If one pizza needs $\frac{3}{4}$ of a metre, how many pizzas can he wrap?

 Challenge

RIBBON

| Blue: 13 metres on each roll | Yellow: 10 metres on each roll | Green: 9 metres on each roll | White: 11 metres on each roll |

The art class are ordering ribbon to use in a craft project. They do not mind what colour the ribbon is. They need to cut the ribbon into pieces that are $\frac{2}{3}$ metre long for their project.

a) How many pieces of ribbon $\frac{2}{3}$ metre long could be cut for each colour?

b) Which colour should they order if they do not want to waste any ribbon? Explain your answer.

1 Convert the following fractions into both a decimal and a percentage. One has been done for you.

a)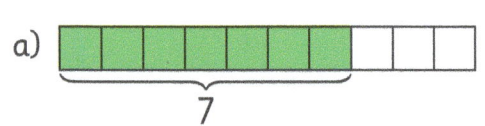

$\frac{7}{10}$ = 0·7 = 70%

b)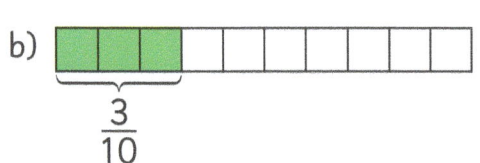

$\frac{3}{10}$ = ⬚ · ⬚ = ⬚ %

c)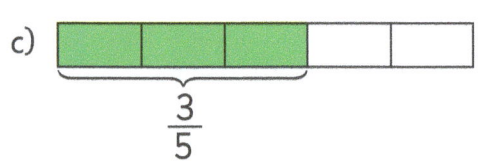

$\frac{3}{5}$ = ⬚ · ⬚ = ⬚ %

d)

$\frac{1}{20}$ = ⬚ · ⬚ = ⬚ %

e)

$\frac{9}{20}$ = ⬚ · ⬚ = ⬚ %

2 Convert each decimal fraction into a fraction in its simplest form, and a percentage. One has been done for you.

a) 0·25 = $\frac{25}{100}$ = $\frac{1}{4}$ = 25 %

b) 0·75 = $\frac{\quad}{\quad}$ = $\frac{\quad}{\quad}$ = ⬚ %

c) 0·32 = $\frac{\quad}{\quad}$ = $\frac{\quad}{\quad}$ = ⬚ %

d) $\boxed{0.15} = \dfrac{\square}{\square} = \dfrac{\square}{\square} = \boxed{}$ %

---

3 Nico and Callum are converting 0·4 into a fraction in its simplest form.

Nico says: I think it is $\dfrac{2}{5}$ and Callum says: No, it is $\dfrac{1}{25}$.

Who do you agree with? Explain your answer.

★ **Challenge**

Sienna has made up clues for an unknown fraction:

- The numerator and the denominator each have exactly two digits.
- The numerator has a 1 in it.
- The denominator has a 4 in it.
- When expressed as a percentage it is less than 45%.

Can you find 4 different possibilities for Sienna's fraction, each with a different denominator?

$\dfrac{\square}{\square} \qquad \dfrac{\square}{\square} \qquad \dfrac{\square}{\square} \qquad \dfrac{\square}{\square}$

1 Use the bar models to work out the following:

a) $\frac{3}{5}$ of 1650

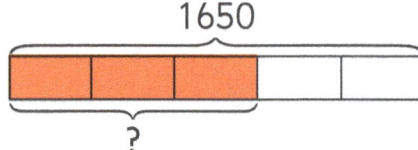

b) 60% of 4500

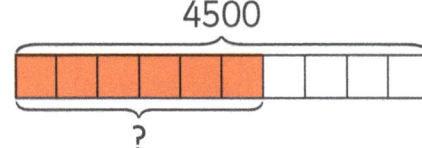

2 Make up a word problem for each of these bar models then solve the problem.

a)

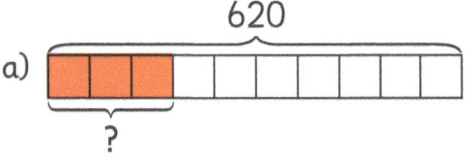

b)

96

?

3 A school is aiming to collect £2400 to pay for a sports trip to France. 30% of this total has been raised so far. How much money still needs to be collected to reach the total?

★ **Challenge**

Some students are testing their fitness in the school health suite. They noted their results in this table but some of the totals got smudged. Work out each student's total and convert each to a percentage.

	FITNESS TEST RESULTS			
	TEST 1	TEST 2	TEST 1 + TEST 2 TOTAL	FINAL %
Aiden	$\frac{14}{20}$	60%	$\frac{130}{200}$	65%
Hamish	$\frac{4}{5}$	$\frac{39}{50}$	$\frac{}{200}$	
Ruby	64%	$\frac{7}{10}$	$\frac{}{200}$	
Abbie	42%	$\frac{11}{25}$	$\frac{}{200}$	

1   Aaron needs to order ink for his printer. He sees these three options on the internet.

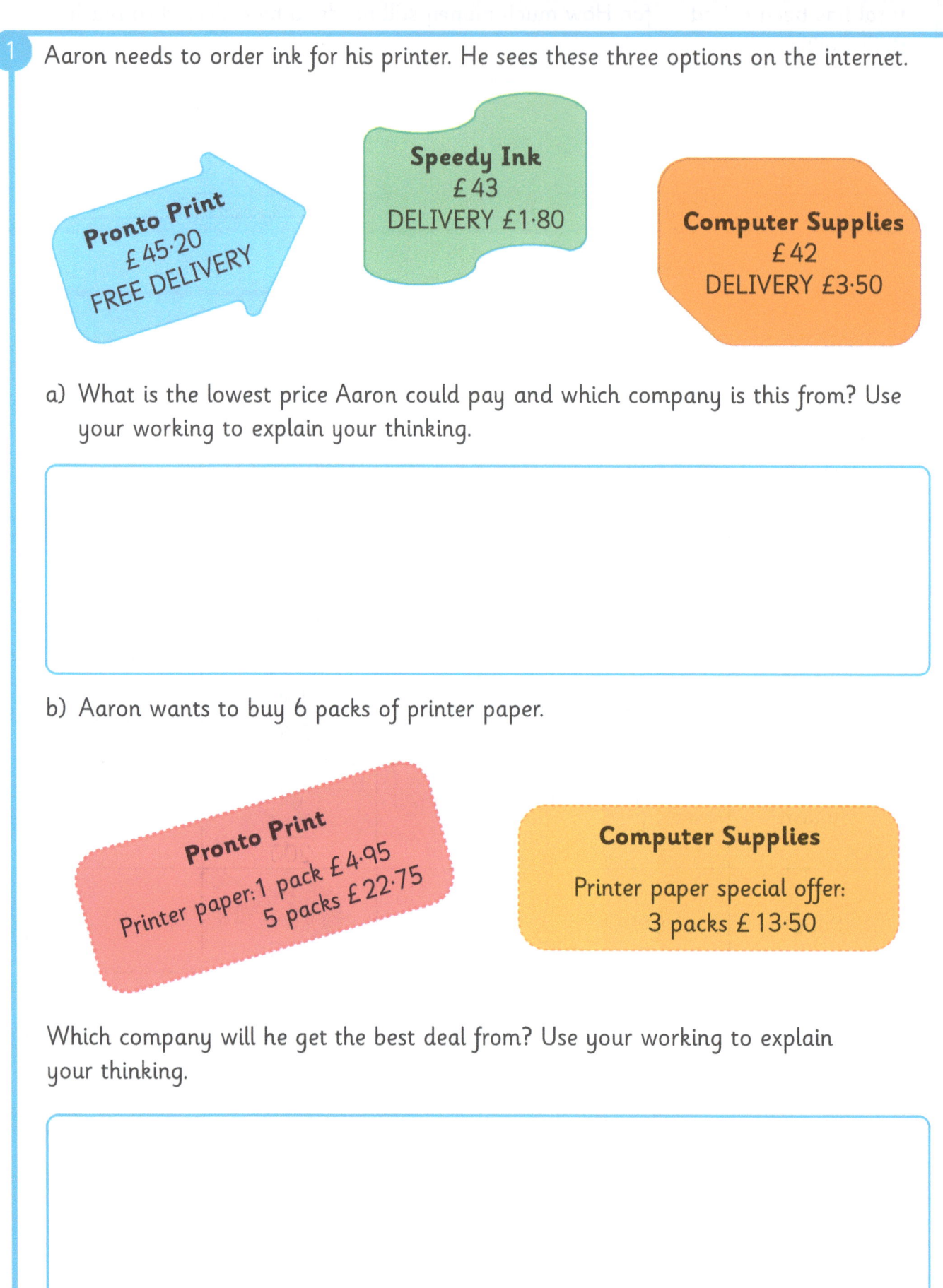

Pronto Print
£45·20
FREE DELIVERY

Speedy Ink
£43
DELIVERY £1·80

Computer Supplies
£42
DELIVERY £3·50

a) What is the lowest price Aaron could pay and which company is this from? Use your working to explain your thinking.

b) Aaron wants to buy 6 packs of printer paper.

Pronto Print
Printer paper: 1 pack £4·95
5 packs £22·75

Computer Supplies
Printer paper special offer:
3 packs £13·50

Which company will he get the best deal from? Use your working to explain your thinking.

2 Grace is planning activities for a school holiday. She says: "I will go swimming on Monday and Tuesday. On Wednesday I will go for a swim then go to the gym."

One swim	£3·60
One swim + gym session	£5·80
Special ticket: three swims + two gym sessions	£14

Should Grace buy a Special ticket? Explain your answer.

★ Challenge

S2 did some baking for a cake stall at a coffee morning. They baked 40 cupcakes to sell at £1·50 each and 35 flapjacks to sell at £1·20 each. They also made 60 chocolate brownies.

a) S2 would like to raise at least £200 from the sale of their baking.
How much do you think they should charge for a chocolate brownie?

b) Riley brings along a cheesecake that can be cut into 12 large slices or 20 small slices. Riley aims to raise at least £20 from the sale of his cheesecake.

How do you think he should cut the cheesecake up and what should he charge for each slice?

1 The table shows the budget plan for a Dance Club over a 4-week period.

	CREDIT	DEBIT
Income (£320 per week)	£1280	
Hall rent (£75 per week)		£300
Teacher's fee (£80 per week)		£320
Electricity (£45 per week)		£180
Heating (£40 per week)		£160
TOTALS	£1280	£960

a) How much does the dance club have left after all the bills have been paid?

b) If the dance club decides to get a mobile phone that costs £4·50 a week, how much will they have left at the end of the 4-week period?

2 Mr and Mrs Willis earn £1650 a month from their gardening business. Their mortgage is £800 a month and they spend £65 a week on food. How much do they have left over to spend on other things each month?

Enter
↓

1	2	3	4
£250	lose £50	add £100	£10 less
5 lose £15	6 £80 more	7 spend £20	8 £60 more
9 plus £50	10 double your money	11 lose £80	12 add £75
13 minus £80	14 plus £95	15 £60 less	16 £85 more

Exit

Square	Add	Subtract	Total
1			£250

Find your way through this maze by going from one square to the next through the gaps. You may only visit a square once. Keep a note of your money total as you go, using the table to help you.

a) What is your money total on your first attempt?

b) Can you find a route that gives you a higher money total?

c) Now try to find a route that gives you a lower money total.

7.3 Profit and loss

1 Poppy bought a skateboard that cost £75, a safety helmet that cost £32 and knee and elbow pads that cost £26. If she sold them all for a total of £120, did she make a profit or a loss? How much was the profit or loss?

2 a) Hudson's mum bought a second-hand caravan for £4560. She wanted to decorate the caravan, then sell it. She spent £420 on material to make new curtains and £390 on new flooring. How much did she spend altogether?

b) Hudson's mum sold the caravan and made a profit of £1130. How much did she sell it for?

3 A craft club made 180 calendars to sell at a Christmas fair. The total cost of making the calendars was £215. They sold 120 calendars at £4 each before Christmas. After Christmas, the remaining 60 calendars were sold at a reduced price of £1·50 each.

What was the total profit they made from selling the calendars?

The Parent Council organised a raffle to raise funds for sports equipment. There were three prizes:

1st prize: £90 voucher
2nd prize: £50 voucher
3rd prize: £30 voucher

The printing of tickets and posters for the raffle cost £40.

The Parent Council plan to sell 140 tickets.

a) Calculate how much each ticket would cost if, after the vouchers had been purchased and the printing paid for, there was no profit. This is called the 'break-even' price.

b) The profit for the raffle was exactly the same as the costs of the prizes and printing. What would be the cost of each ticket to create this profit?

c) If 170 tickets are sold, the prizes remain the same and the cost of printing is still £40, how much would each ticket need to cost to make a profit of £385?

7.4 Discounts

1 Erin wants to buy a new tennis racquet that costs £55.

A sports shop has a sale giving 20% off all tennis equipment. What is the discounted price?

20% off

2

Buy one, get one half-price

COMFY CLOTHING
AMAZING OFFERS

20% off all clothing

Jackson wants to buy two new hoodies. He sees some that cost £18 each in a shop with an offer of "buy one get one half-price". The store is also offering 20% off all clothing. Jackson can only choose one of the offers.

a) Which offer will save Jackson the most money?

b) How much does Jackson save on the full price of two hoodies?

3 This leaflet shows the ticket prices for entry to a museum.

Adult ticket	£15
Child ticket	£10
Family ticket A (2 adults, 2 children)	£46·50
Family ticket B (1 adult, 3 children)	£44

The Kaur family need to buy tickets for 2 adults and 4 children.

Work out the cheapest way for them to do this.

★ Challenge

The drama teacher is planning a class outing to either the cinema or the theatre. Ticket prices and discount offers for each are shown here.

ORION CINEMA

Adult ticket	£10
Child ticket	£7

Loyalty card holders 10% discount

STAGES THEATRE GROUP
present
"SCHOOL ROCKS"

Adult ticket	£12
Child ticket	£10

1 free child ticket for every 10 bought
School groups 20% discount

The teacher has a loyalty card for the cinema.

30 children and 5 adults will be going on the outing. Will it be cheaper to buy tickets for the cinema or the theatre? Use your working to explain your thinking.

Cinema	Theatre

1 Mr Higgins wants to buy an electric bike that costs £650.

CYCLE SHACK

Electric bike	**£ 650**
Finance availabe:	
Deposit	£ 100
12 monthly payments of	£ 55

If Mr Higgins buys the bike using the finance available, how much will he pay altogether?

2 Mrs Ahmed is going to buy this games console. She wants to know how much she will save if she pays the cash price instead of paying in instalments.

GAMING DIRECT

Games console cash price...................................£ 475

Or pay in instalments:
Deposit...£ 90

6 monthly payments of....................................£ 70

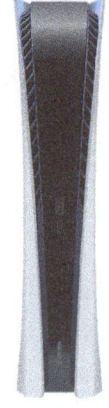

How much money will Mrs Ahmed save if she pays the cash price?

3

SOFA CENTRAL

Hire purchase available at no extra cost.
No deposit required.

Annabel sees this special offer and decides to buy a sofa that costs £1195. She chooses to make six equal monthly payments of £175. How much will Annabel have to pay in the seventh month to pay the sofa off?

★ Challenge

Hire purchase is available on a family holiday to EuroWorld.

EUROWORLD

Family holiday for 2 adults + 2 children

Deposit **£800 + 18 monthly payments** of **£240**

Special Offer
20% off the hire-purchase price for full payment when booking.

The McKay family decide to take advantage of the special offer and pay for the holiday when they book.

a) How much do they pay for their holiday? Show all of your working.

b) How much do the McKay family save by paying for their holiday when they book?

1 Lucas and his family are travelling from Glasgow to Inverness by train. The journey takes 3 hours and 10 minutes. If they leave Glasgow at 11·50 am, what time will they arrive in Inverness?

2 A school trip to London is being planned. The organisers need to decide if they will travel by bus, train or aeroplane.

- Going by bus takes 12 hours and 20 minutes
- Going by train takes 6 hours and 45 minutes
- Going by aeroplane takes 1 hour 25 minutes

a) If they decide to go by bus and leave at 6 am, what time will they arrive in London?

b) A flight arrives in London at 3·55 pm. What time does it leave?

c) If they want to arrive in London by 3 pm and decide to go by train, what is the latest time they can catch a train?

3 Two trains travel on the same route between Montrose and Glasgow:

First train: leaves Montrose at 8·40 am and arrives in Glasgow at 10·58 am

Second train: leaves Montrose at 1·12 pm and arrives in Glasgow at 3·04 pm

Which train do you think makes the most stops? Explain your thinking.

★ Challenge

Some students took part in an outdoor activities event. Nicki and Shelley were in one group and Joe and Craig were in a different group.

This is Nicki's report of their activities:

We left the outdoor centre at 10 am. We started with a 10 minute walk to the loch. Then we got in and swam to the other side. This took us 25 minutes. When we got out we had 15 minutes to get dried and changed before we got on bikes and cycled for 45 minutes to the barbecue area.

Joe wrote this to say what his group did:

We got off the bus at 9·45 am. 15 minutes later we got into kayaks. We paddled across the loch for 50 minutes. We took 10 minutes to get out of the kayaks and get ready for our run. We ran for 20 minutes before we realised we were lost. We spent 5 minutes getting back to the track then we ran for 15 minutes. We were glad to arrive at the barbecue area.

Which group arrived at the barbecue area first? Show your working.

1 S3 are at an outdoor activities park. They see this list of activities that they can take part in.

Nature walk	starts at 10·15 am	lasts 1 hour 20 minutes
Kayaking	starts at 10·30 am	lasts 2 hours 30 minutes
Mountain bike tour	starts at 10·40 am	lasts 3 hours 15 minutes

a) The start of the nature walk is delayed by 10 minutes. What time will it finish now?

b) Kai and Niamh need to catch the bus home at 1·45 pm. Can they go on the mountain bike tour? Explain your answer.

2 As a birthday treat, Layla is going on the Seabirds and Seals Cruise.

Layla is on the boat for 15 minutes before it leaves and 10 minutes after it returns. How long is she on the boat for altogether?

RIVER FORTH BOAT TRIPS

Seabirds and Seals Cruise

Leaving at 11·35 am
Returning at 2·10 pm

Four friends are planning a trip to the cinema to see a movie called Cyber Corruption that begins at 2 pm.
They will meet in the town centre and take a bus on route 32 to the cinema.

ORION CINEMA

Special screening of
CYBER CORRUPTION
At 2 pm

BUS ROUTE 32					
Town centre	11·15 am	11·50 am	12·20 pm	1·00 pm	1·20 pm
Swimming pool	11·18 am	–	12·23 pm	–	1·23 pm
Retail park	–	11·55 am	12·25 pm	–	1·25 pm
High school	11·25 am	–	12·30 pm	–	1·30 pm
Park and ride	11·30 am	12·05 pm	12·35 pm	1·12 pm	1·35 pm
Cinema	11·33 am	12·08 pm	12·35 pm	1·15 pm	1·38 pm

Freya says:

When we get off the bus it will take us 12 minutes to walk to the cinema.

Mason says:

We need to allow 10 minutes to pick up our cinema tickets.

Ava says:

I want to go to the kiosk when I get there to get a drink and a snack.

Fred says:

Yes! Let's all go to the kiosk together before we go into the movie.

Which bus do you think they should catch for their cinema trip?
Explain your thinking.

8.3 Investigating ways speed, time and distance can be measured

1 Use the formula **time = distance ÷ speed** to calculate the time taken for each of these journeys:

a) Ella walked 12 miles at 3 mph.

hours

b) An athlete ran 15 km at 5 kph.

hours

c) A bus travelled 350 miles at 50 mph.

hours

d) Rory cycled 28 km at 7 kph.

hours

2 Use the formula **speed = distance ÷ time** to calculate the average speed of these journeys:

a) A lorry travelled 120 miles in 3 hours.

b) A car was driven 120 miles in 2 hours.

c) A helicopter flew 480 km in 4 hours.

d) A swimmer swam 1500 metres in 50 minutes.

3 Use the formula **distance = speed × time** to calculate how far each of the following travelled:

a) A bike was ridden at a speed of 20 kph for 2 hours.

b) A car was driven at a speed of 60 mph for 3 hours.

c) A bird flew for 5 hours at a speed of 25 mph.

d) A train travelled for 4 hours at a speed of 125 kph.

4 Choose the correct formula to work out each of these.

a) A motorcycle was driven at 45 kph for 2 hours. How far was the journey?

b) A drone flew 180 metres at a speed of 9 metres per second. How long did the drone's journey take?

★ **Challenge**

A car travelled for 90 minutes at a speed of 50 mph. How far did the car travel? Show your working.

1 Some students are taking part in a construction challenge. They time each other building a tower using 25 modelling bricks

Here are their times, shown on a stopwatch.

Tom's time:	Leon's time:	Keira's time:	Eve's time:
03:21 32	02:50 85	03:09 02	02:58 10

a) Who took the longest time to build their tower?

b) Who completed the task in the shortest time?

c) Write the students' names in order of finishing the task, from quickest to slowest.

2 Three runners in a 2 km fun run wrote down their **target** finish times before they started running.

Eleanor	15 minutes and 30 seconds
Alisha	12 minutes
Neda	13 minutes

These stopwatches show the **actual** time each runner took in the fun run:

Eleanor's time:

13:15 07

Alisha's time:

13:28 68

Neda's time:

12:55 00

a) Who took less time to run the 2 km than their target time?

b) Who in this group of three runners was first to cross the finishing line?

c) Neda was very happy to beat her target time. How much did she beat her target time by?

★ Challenge

Fill in the blanks to complete the table. One has been done for you.

START TIME	FINISH TIME	TIME ELAPSED
04:07 16	08:19 23	4 minutes, 12 seconds, 7 centiseconds
12:15 28	15:23 32	
06:31 15		8 minutes, 15 seconds, 21 centiseconds
	24:08 49	3 minutes, 5 seconds, 18 centiseconds
		12 minutes, 50 seconds, 3 centiseconds

8.5 Converting between units of time

1 Convert the following from seconds to minutes:

a) 180 seconds

b) 360 seconds

c) 720 seconds

d) 1500 seconds

2 Convert the following from minutes to hours and minutes:

a) 120 minutes

120 minutes = _____ hours _____ minutes

b) 150 minutes

150 minutes = _____ hours _____ minutes

c) 270 minutes

270 minutes = _____ hours _____ minutes

d) 690 minutes

690 minutes = _____ hours _____ minutes

3 Convert the following from hours to days:

a) 48 hours

b) 120 hours

c) 216 hours

d) 600 hours

⭐ **Challenge**

a) Sort these times into **ascending order**, starting with the shortest time:

0·5 hours   72 minutes   6,000 seconds   95 minutes   2·75 hours   3,240 seconds

b) Sort these times into **descending order**, starting with the longest time:

3 years   116 weeks   42 months   260 weeks   2·5 years   35 months

1 A laser measuring device was used to find measurements in the school outdoor area.

The device gave the measurements in metres, but they need to be converted into centimetres.

Convert each of these lengths into centimetres and write your answers in the blank boxes above the measurements.

SCHOOL OUTDOOR AREA

Trim trail

[   ] by [   ]

4·65 m   by   2·15 m

Recycling bins

[   ] by [   ]

3·2 m by 2·5 m

Greenhouse

[   ] by [   ]

2·4 m by 2 m

Outdoor gym

[   ] by [   ]

6·79 m   by   5·28 m

Vegetable garden

[   ] by [   ]

3·59 m by 3·10 m

Bike racks

[   ] by [   ]

2·1 m by 3·04 m

Covered shelter

[   ] by [   ]

4·92 m by 3·11 m

2 Some students were finding out about bridges. They wrote down the lengths in metres of these two bridges in Scotland. Convert each of these lengths into kilometres.

a) Forth Road Bridge: 2512 metres

b) Queensferry Crossing: 2700 metres

3 The following list gives the diameter in millimetres of coins with a value of less than £1.

Convert each diameter into centimetres.

Coin	Diameter in mm	Diameter in cm
1p	20·3 mm	
2p	25·9 mm	
5p	18·0 mm	
10p	24·5 mm	
20p	21·4 mm	
50p	27·3 mm	

A sunflower was measured three times a week for five weeks. This table shows how much it had grown each time it was measured.

	Monday	Wednesday	Friday
Week 1	8·4 cm	75 mm	9·8 cm
Week 2	109 mm	10·2 cm	4 cm
Week 3	18·4 cm	42 mm	8·7 cm
Week 4	2·6 cm	12·5 cm	0·34 m
Week 5	45 mm	16 mm	26 mm

a) At the beginning of week 1, the sunflower measured 3·1 cm. Complete this table to show the height of the sunflower after each measurement. The first week has been filled in for you.

	Monday	Wednesday	Friday
Week 1	3·1 cm + 8·4 cm = 11·5 cm	11·5 cm + 75 mm = 19 cm	19 cm + 9·8 cm = 28·8 cm
Week 2			
Week 3			
Week 4			
Week 5			

b) The gardener thinks it grew most in Week 5. Do you agree? Explain your thinking.

9.2 Estimating and measuring mass

1 Angus and Lara are making a traybake to serve at a school open day. They weigh the ingredients using digital scales that show mass in kilograms.

Write the mass of each ingredient in grams.

a) Butter

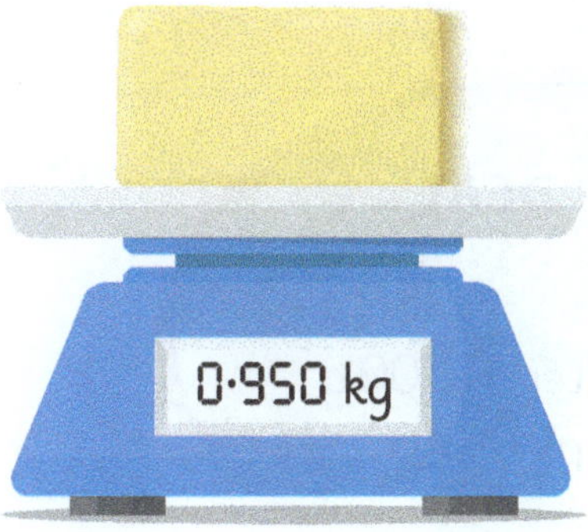

0·950 kg

b) Golden syrup

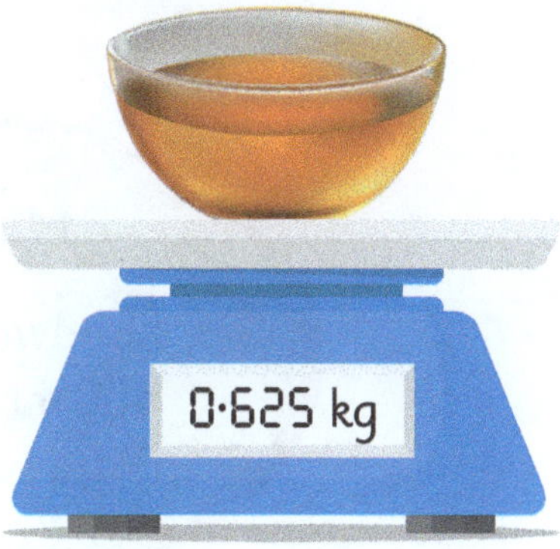

0·625 kg

c) Brown sugar

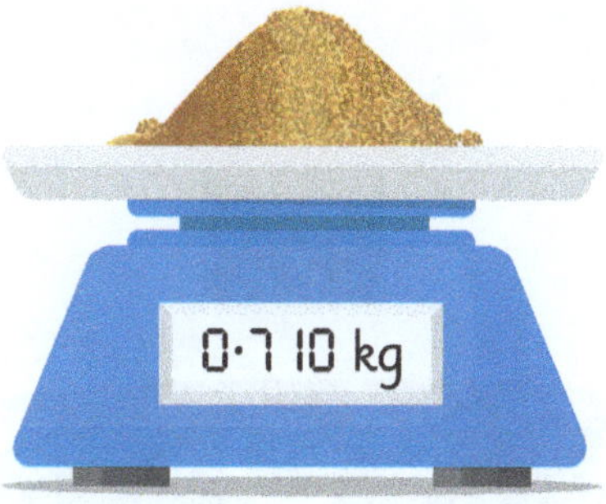

0·710 kg

d) Oats

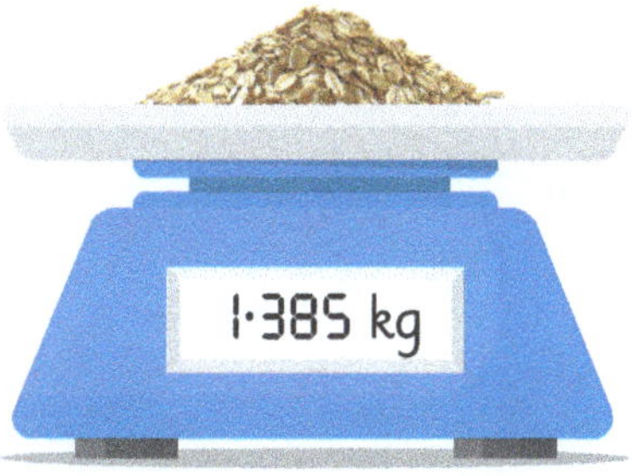

1·385 kg

e) Dried fruit

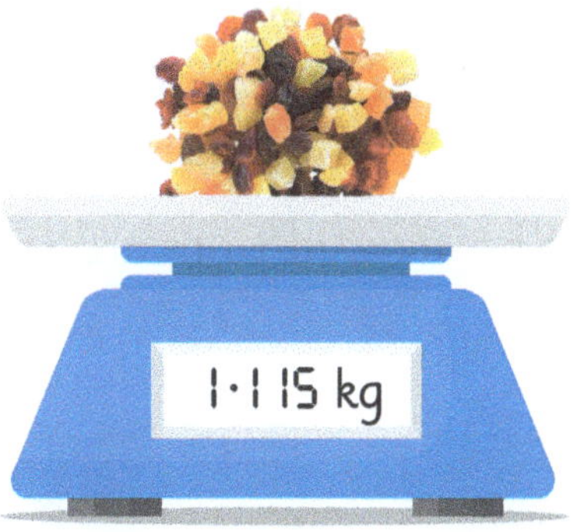

1·115 kg

f) Milk chocolate

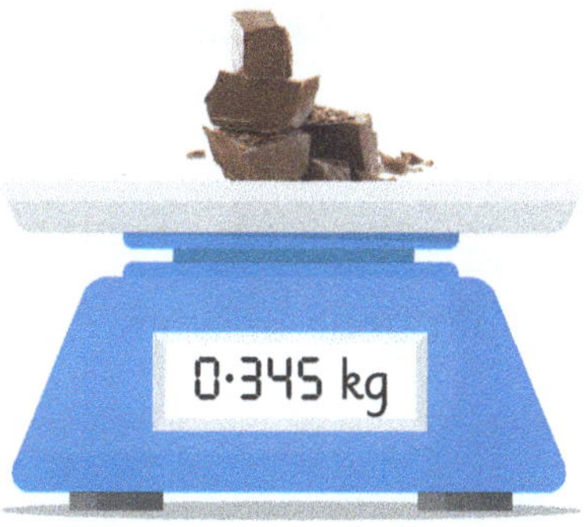

0·345 kg

2 Write these in ascending order, starting with the lightest:

2385 g   1·95 kg   2·55 kg   985 g   0·649 kg

3 Mrs Pearson checks the mass of her hand luggage before she leaves for the airport.

She knows she must not take more than 7·5 kg with her onto the flight.

Mrs Pearson's scales are difficult to read, but she estimates that they are showing around 6 kg for the mass of her hand luggage.

MEAL DEAL

Sandwich	280 g
500 ml water	500 g
Potato crisps	40 g

At the airport, Mrs Pearson buys a meal deal to take onto the aeroplane with her. She also buys a hardback book that has a mass of 850 g. She puts the meal deal and the book into her hand luggage.

Do you think Mrs Pearson will be allowed to board the aeroplane?
Explain your answer.

At an autumn fair there is a pumpkin growing competition.

Points are given according to the mass of the pumpkin, with 1 point being awarded for every gram. Here are the results of the competition:

PUMPKIN GROWING COMPETITION

First place	4050 points
Second place	4007 points
Third place	3960 points

a) Write down the mass of each prize-winning pumpkin in kilograms.

b) One of the entrants goes to the judge to complain, saying:

> I should have 3rd prize because my pumpkin weighs 3 kilograms and 98 grams.

How do you think the judge will explain the results to this person?

1 Calculate the area of each rectangle. The first one has been partitioned for you.

a)

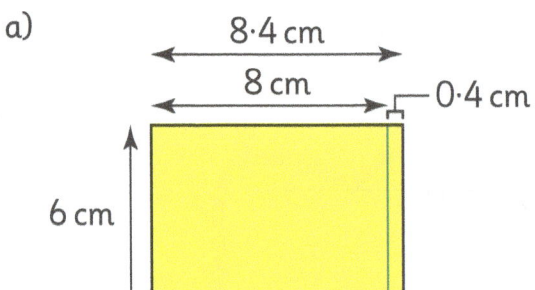

b)

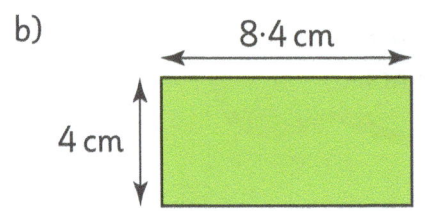

c)

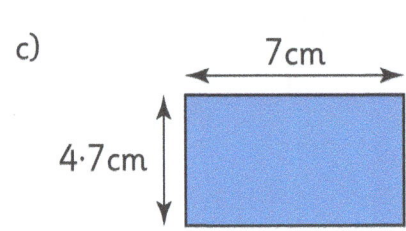

2 Find the missing length for each window.

a)

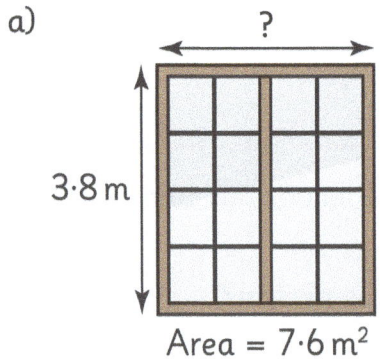

Area = 7·6 m²

b)

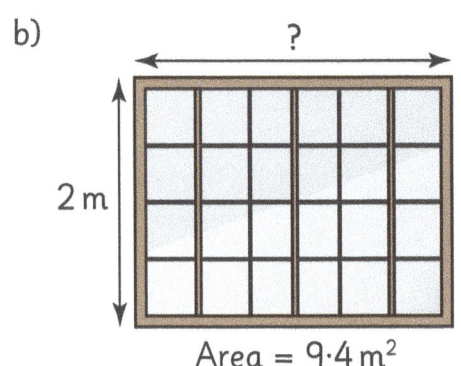

Area = 9·4 m²

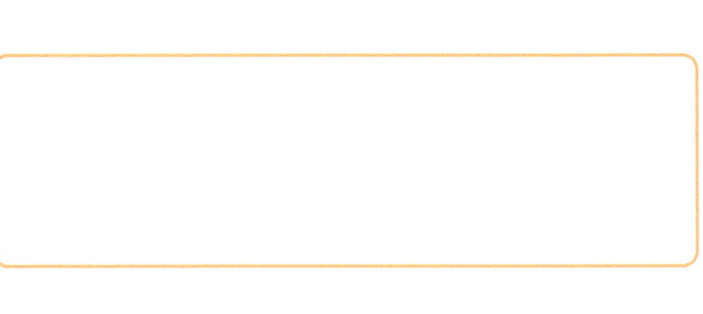

3 Simon is choosing a rug for his room, which has an area of 6 m². He wants the rug to cover as much of the floor as possible. Tick the rug he should choose. Show your working.

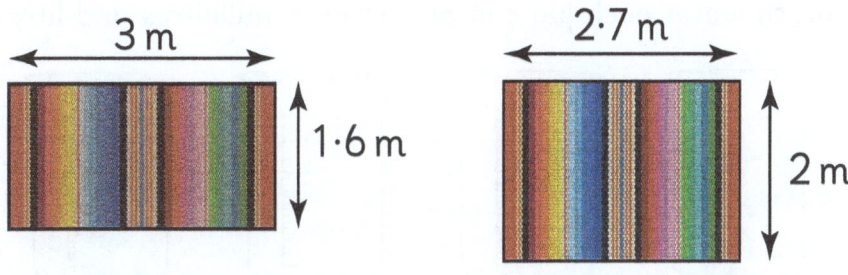

3 m, 1·6 m

2·7 m, 2 m

★ **Challenge**

Taylor and Lucie have each been asked to draw a rectangle with an area of 10·26 cm².

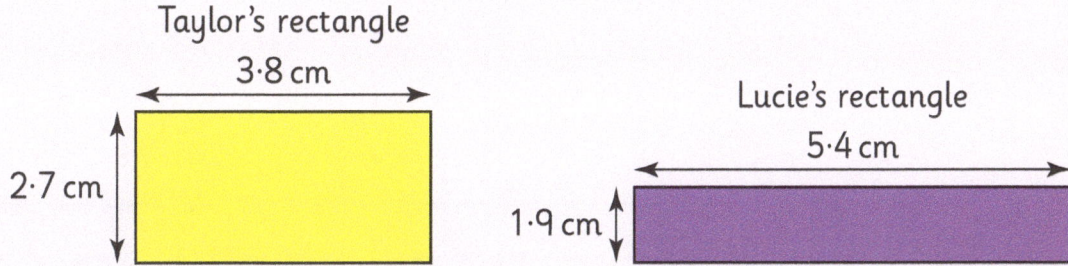

Taylor's rectangle
3·8 cm
2·7 cm

Lucie's rectangle
5·4 cm
1·9 cm

Have they done this correctly? Explain your thinking.

9.4 Estimating and measuring capacity

1

a) Some students have poured water into jugs for a science experiment. Write how much water each jug contains in both millilitres and litres.

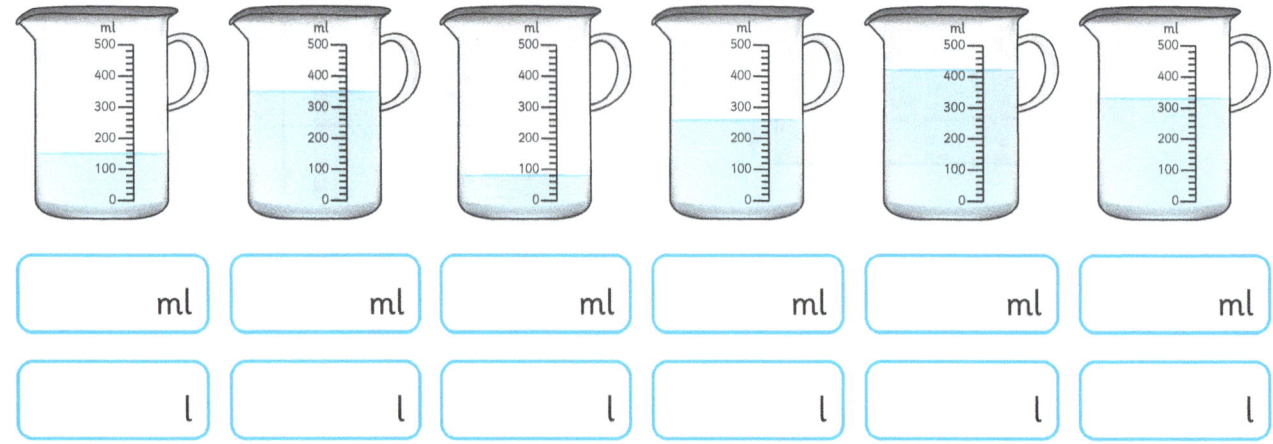

| ml | ml | ml | ml | ml | ml |

| l | l | l | l | l | l |

b) The experiment needs the students to pour exactly 1 litre of water into a bowl. Can they do this using the amounts they have? Explain your answer.

2 Draw a line on each jug to show the given volume of liquid. Shade in the liquid below the line.

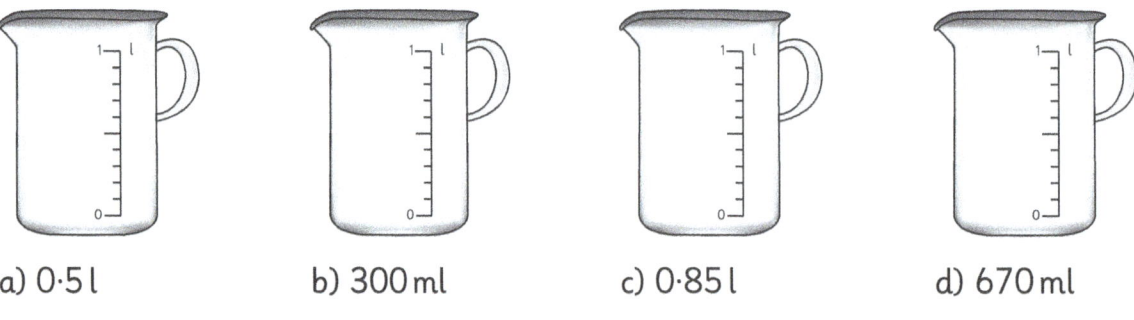

a) 0·5 l   b) 300 ml   c) 0·85 l   d) 670 ml

3 A bottle contains 2 litres of milk. Two glasses of milk are poured from the bottle, one containing 200 ml and the other containing 350 ml.

ml

l

How much milk will be left in the bottle? Give your answer in millilitres and also in litres.

★ **Challenge**

You have 1·5 litres of orange squash to share equally between three containers: a jug, a bottle and a carton.

Jug

Bottle

Carton

Each container holds at least 300 ml of liquid. The carton holds more than the bottle but less than the jug.

Using this information, can you find two different ways to share the orange squash?

9.5 Estimating imperial measurements

1 The average length of a jaguar from nose to tail is 9 feet. Compare this to the other big cats shown here and use it to estimate the average length of each one in feet.

Tiger

Jaguar 9 feet

Leopard

Lynx

Wildcat

2 This map shows some main
UK airports. The distance between
Edinburgh airport and Inverness airport
is 150 miles. Estimate the following
distances:

SHETLAND ISLES

INVERNESS
ABERDEEN
EDINBURGH
GLASGOW
NEWCASTLE
MANCHESTER
BRISTOL
LONDON
HEATHROW

a) Manchester airport
and Glasgow airport

b) London Heathrow
airport and Aberdeen
airport

c) Newcastle airport and
Bristol airport

d) Inverness airport and
London Heathrow
airport

e) Aberdeen airport and
Inverness airport

f) Manchester airport
and Inverness airport

★ **Challenge**

1 pound = 0·45 kilograms

1 kilogram = 2·2 pounds

4 oranges **or** 4 bananas have a mass of roughly one pound.

Use this to help you to work out the mass of 6 oranges **and** 4 bananas in kilograms.

9.6 Converting imperial measurements

Use the following to convert between metric and imperial measurements:

Length	Mass	Capacity
1 inch = 2·54 cm	1 ounce = 28·3 g	1 fluid ounce = 28·4 ml
1 foot = 30·5 cm	1 kg = 2·2 pounds	1 cup = 240 ml
1 yard = 91·4 cm	1 stone = 6·35 kg	1 litre = 1·76 pints
1 mile = 1·61 km		1 gallon = 4·55 litres

1 Convert:

a) 18 inches to cm

b) 78 kg to stones

c) 15 fluid ounces to ml

d) 12 pints to litres

2 The measurements for a tennis court are given here in feet.

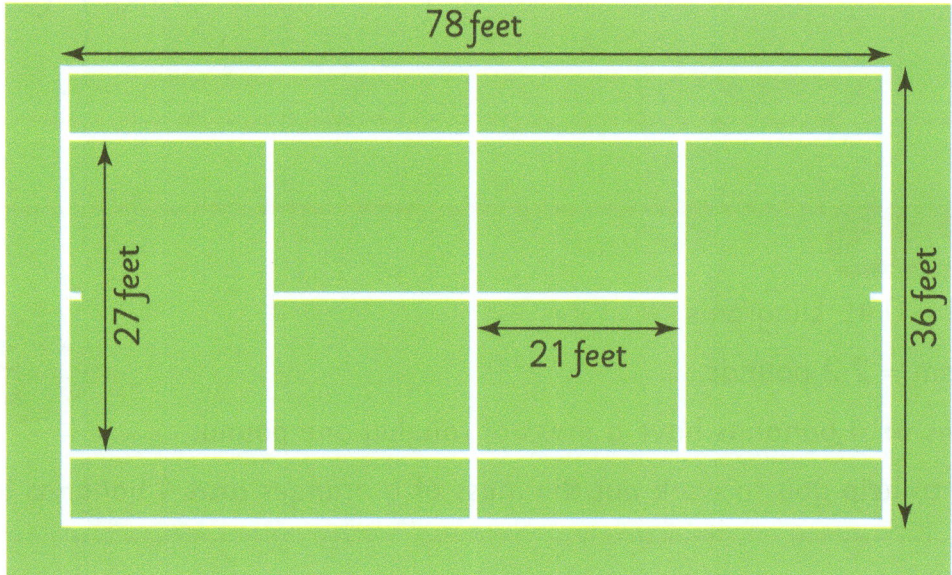

Convert each measurement to both metres and centimetres.

a) 78 feet

b) 36 feet

c) 27 feet

d) 21 feet

★ **Challenge**

Arrange each of these into ascending order, starting with the smallest:

a) 80 feet; 3000 inches; ½ kilometre; 15 yards; 120 metres

b) 2 pounds; ¼ kilogram; ½ stone; 16 ounces; 702 grams

c) 50 fluid ounces; 13 cups; 2 litres; 1 gallon; ½ pint

1 Calculate the perimeter of each of the shapes below:

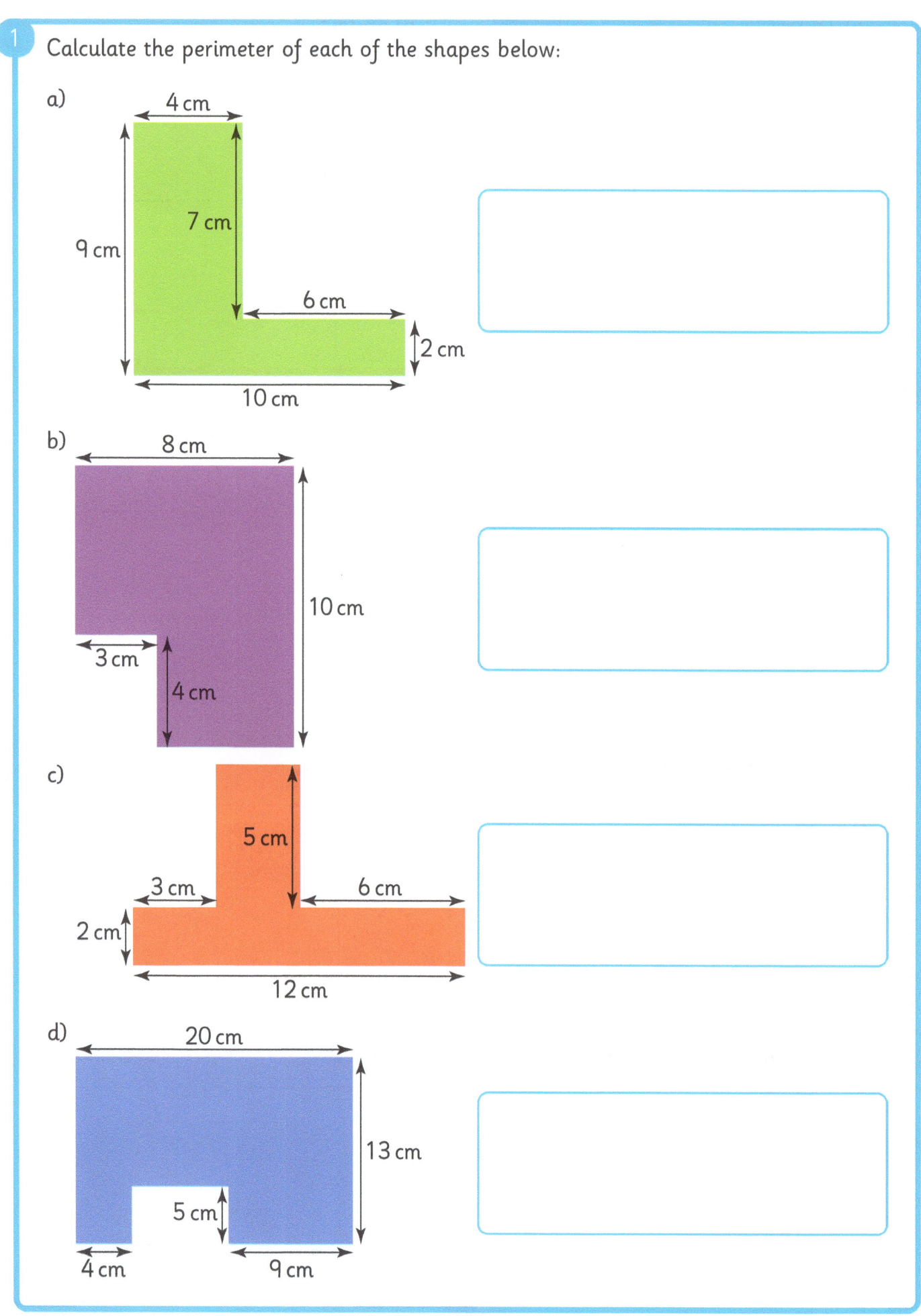

a)
4 cm
7 cm
9 cm
6 cm
2 cm
10 cm

b)
8 cm
10 cm
3 cm
4 cm

c)
5 cm
3 cm
6 cm
2 cm
12 cm

d)
20 cm
13 cm
5 cm
4 cm
9 cm

2 This is a plan of a community garden.

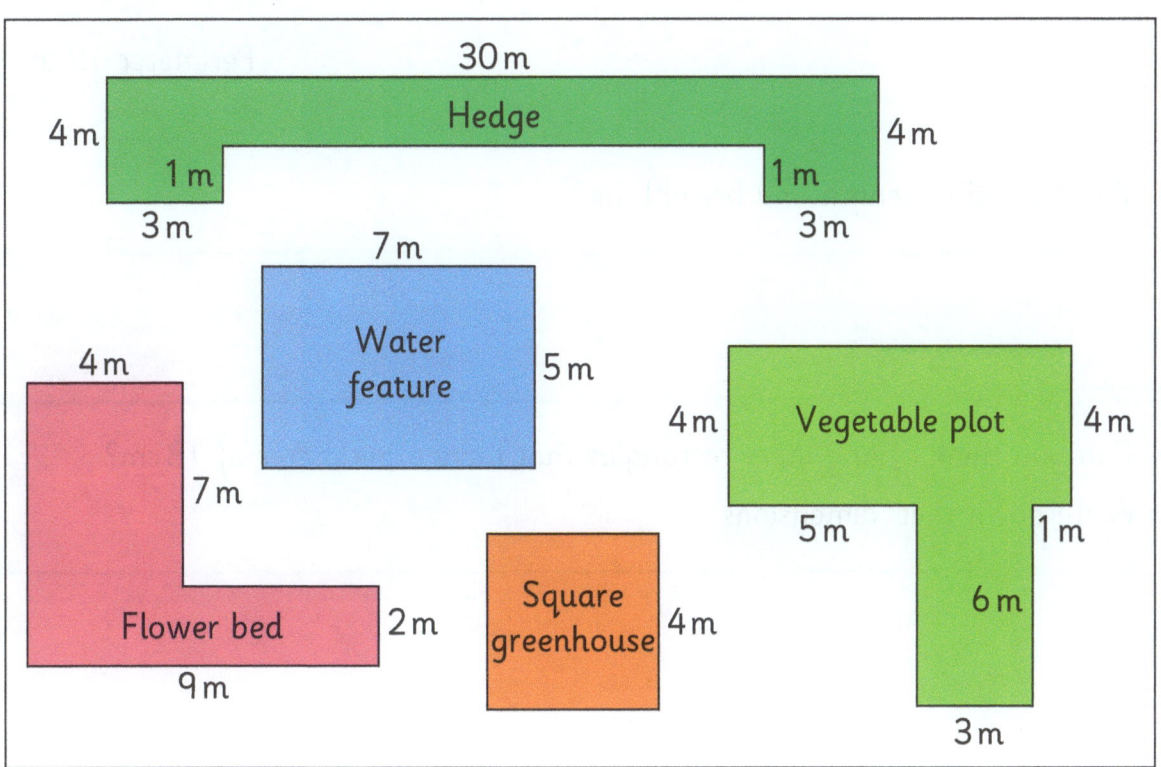

Calculate the perimeter of each area in the garden.

a) Hedge

b) Water feature

c) Greenhouse

d) Flower bed

e) Vegetable plot

3 A rectangle has a perimeter of 18 cm.

Perimeter 18 cm

a) What could its length and breadth be?

b) Can you think of any more rectangles that have a perimeter of 18 cm? Write down their dimensions.

★ **Challenge**

The top of a desk is in the shape of a rectangle. The length of the desk is 45 cm more than its width. If the perimeter of the desk is 6·3 m, calculate its length and width.

1 Calculate the area of each of the following:

a)

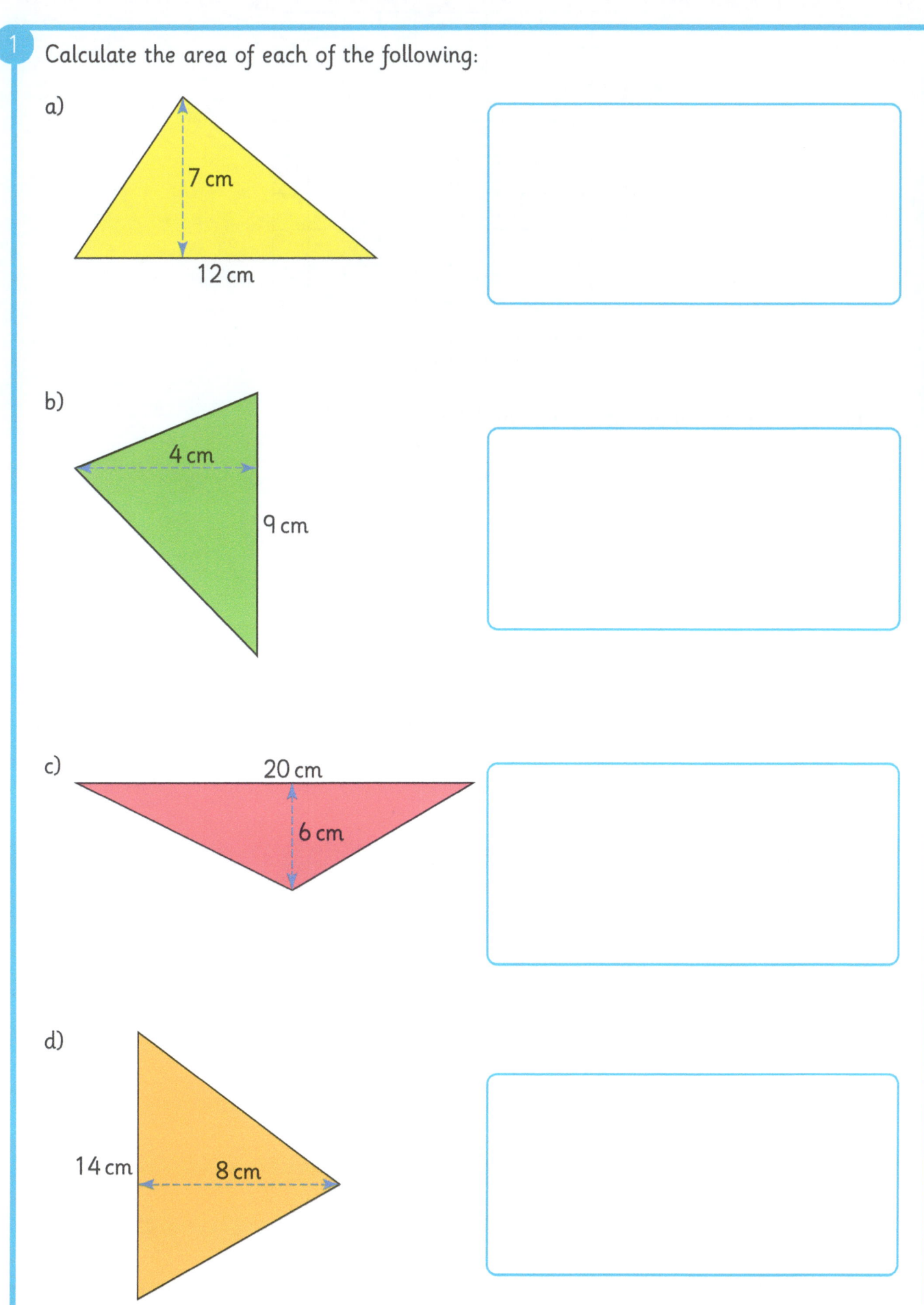

7 cm

12 cm

b)

4 cm

9 cm

c)

20 cm

6 cm

d)

14 cm

8 cm

2 a) Draw a triangle that has an area of 15 cm².

b) Draw a triangle that has an area of 18 cm².

c) Draw a triangle that has an area of 24 cm².

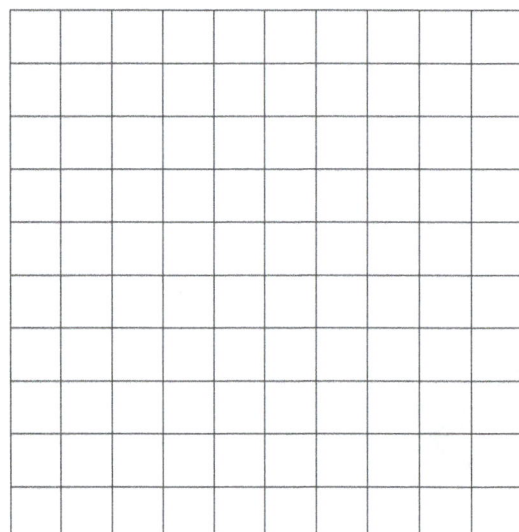

3 These two triangles have the same area. Calculate the missing length, shown by **?** on the diagram.

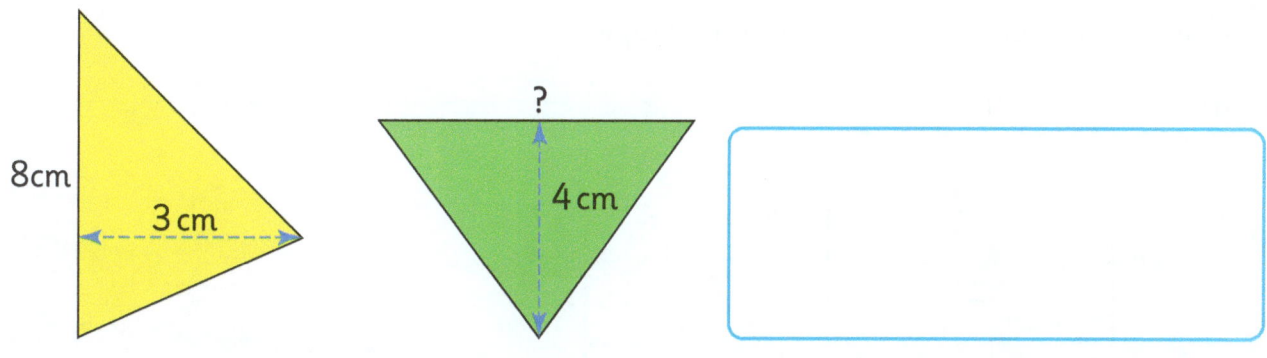

8cm

3 cm

?

4 cm

★ **Challenge**

The area of a triangle is greater than 13 cm² but less than 15 cm². The triangle's length is three times greater than its height. Can you work out what the length and height of this triangle might be?

1 Calculate the area of each composite shape.

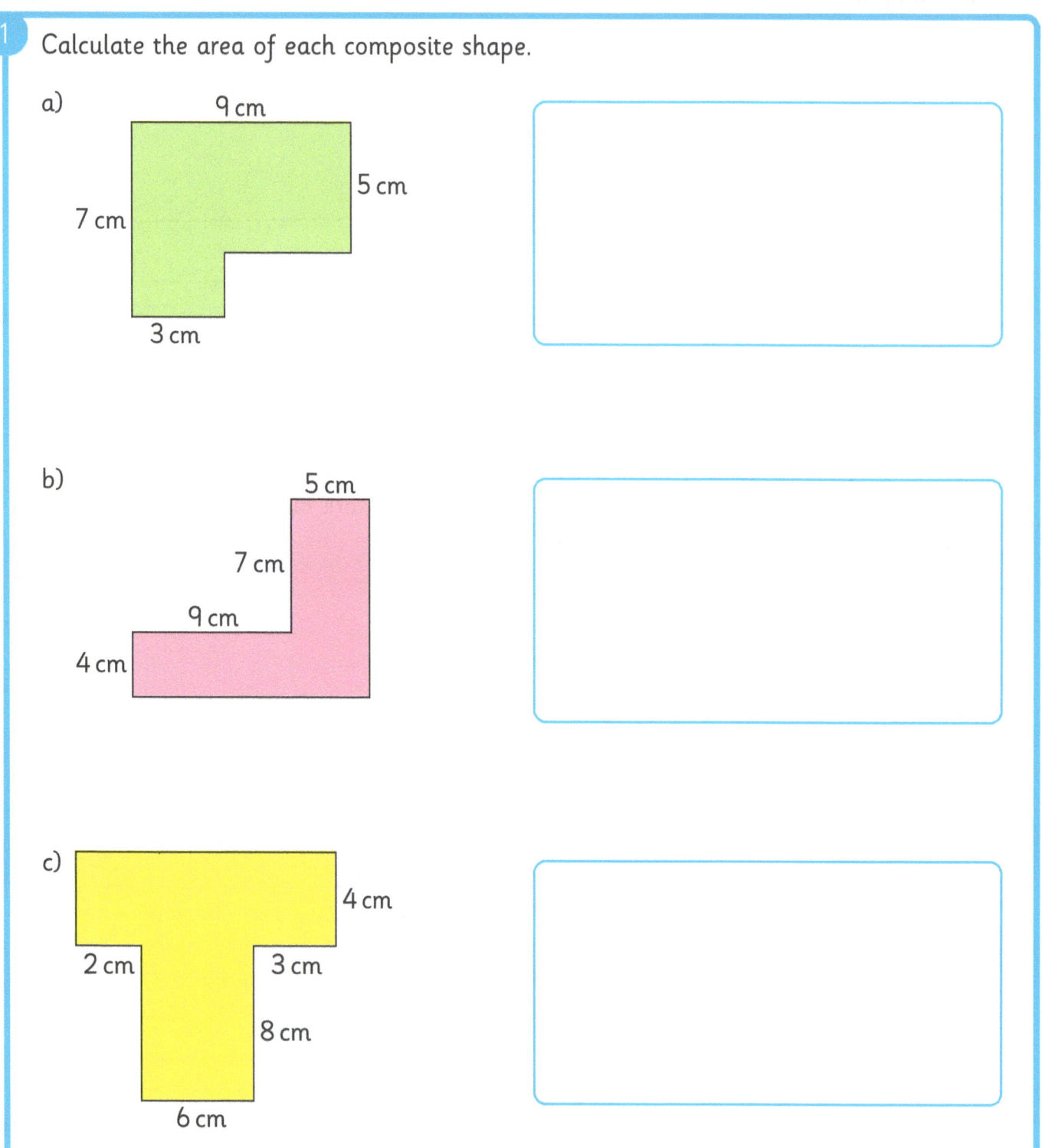

a)

b)

c)

2 Calculate the area of each part of the community garden:

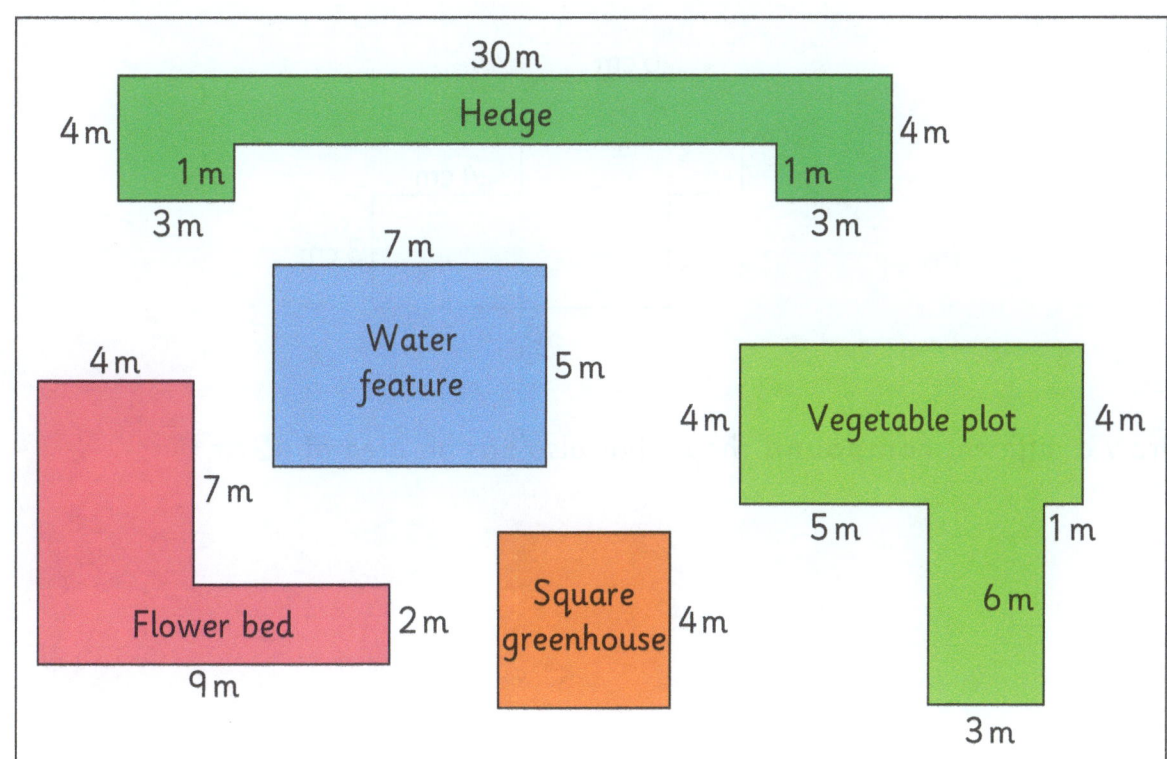

a) Hedge

b) Water feature

c) Greenhouse

d) Flower bed

e) Vegetable plot

3 The area of this compound shape is 42 cm².

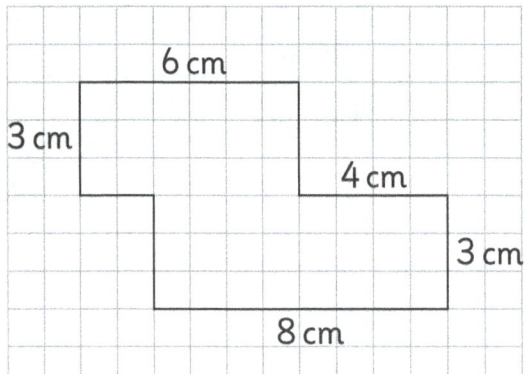

Draw a different **compound** shape that also has an area of 42 cm².

★ Challenge

In this compound shape, the area of A is double the area of B.

What could the length of each side be?

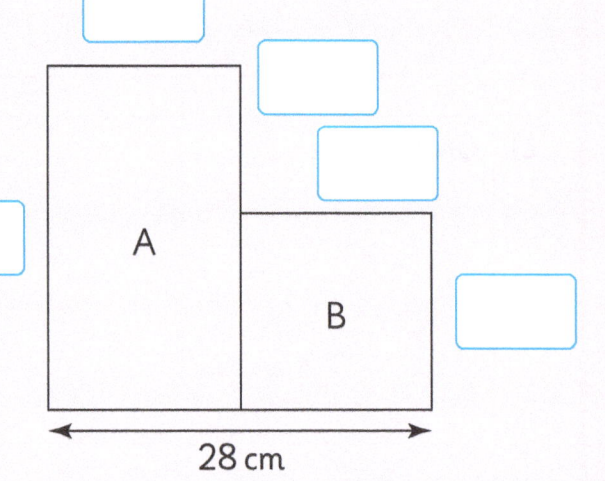

28 cm

1 Calculate the volume of each box. They are not drawn to scale.

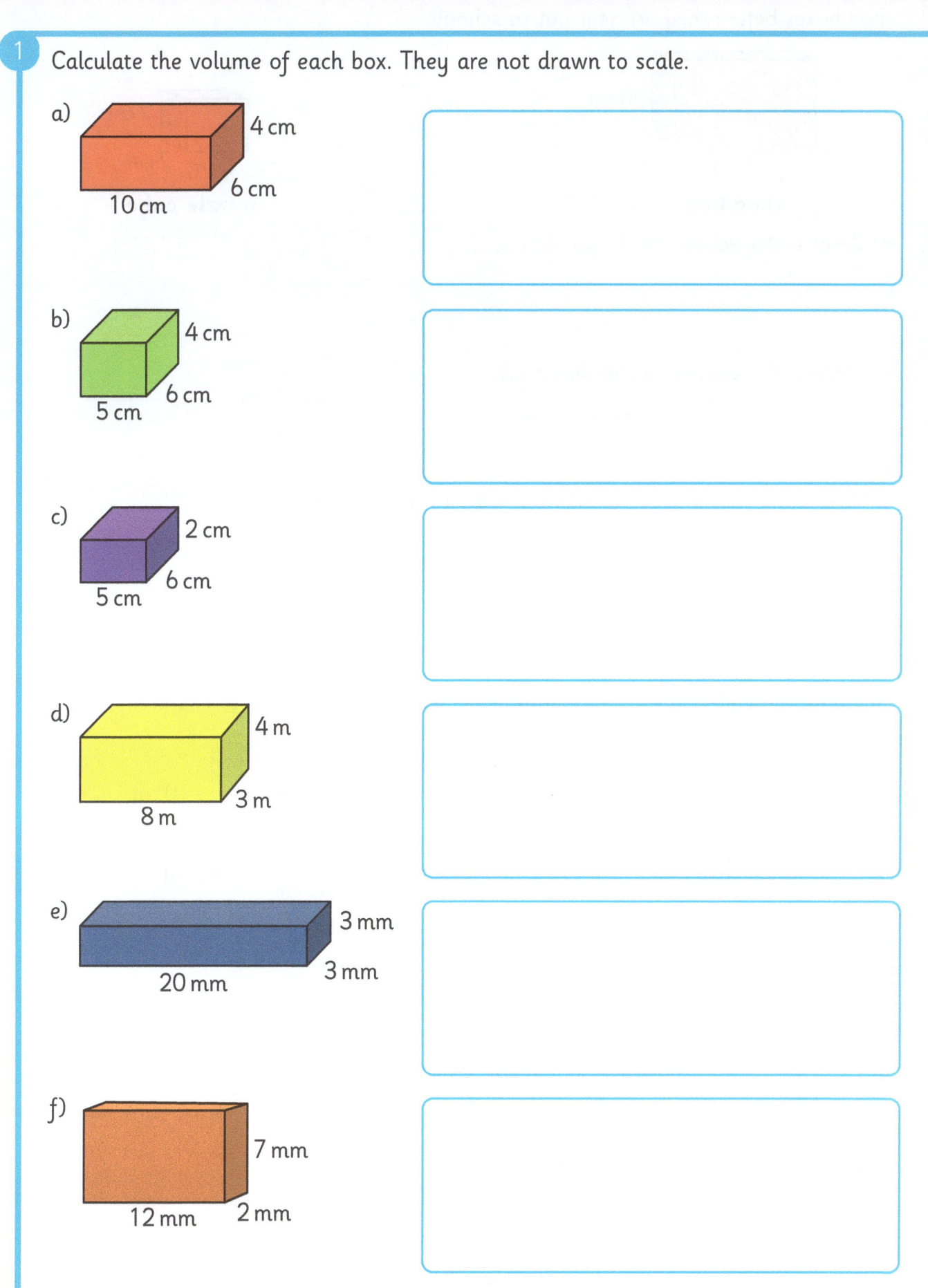

a) 4 cm, 6 cm, 10 cm

b) 4 cm, 6 cm, 5 cm

c) 2 cm, 6 cm, 5 cm

d) 4 m, 3 m, 8 m

e) 3 mm, 3 mm, 20 mm

f) 7 mm, 2 mm, 12 mm

2 A toy company is donating puzzle cubes to local schools. The cubes will be packed in shoe boxes before they are sent out to schools.

20 cm
30 cm   20 cm
shoe box

2 cm
2 cm   2 cm
puzzle cube

a) What is the volume of the puzzle cube?

b) What is the volume of the shoe box?

c) How many puzzle cubes can be packed into a shoe box?

3 Abe and Kim collect these two empty boxes from a supermarket to use in a project.

POTATO CRISPS
60 cm
75 cm   50 cm

WHEAT BISCUITS
110 cm
55 cm   40 cm

Abe says: We need to use the box with the largest volume. That is the wheat biscuits box.

Kim says: I think we should check because the potato crisps box looks larger.

Which box should they choose? Explain your answer.

These three cuboids have equal volume. Work out what the missing lengths could be.

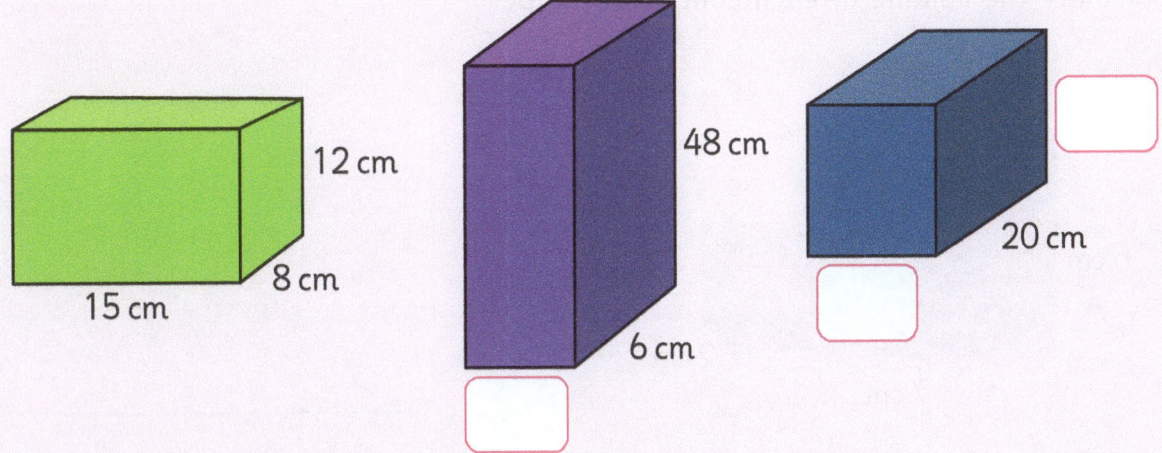

12 cm
8 cm
15 cm
48 cm
6 cm
20 cm

1 Calculate the volume of each composite shape.

a)

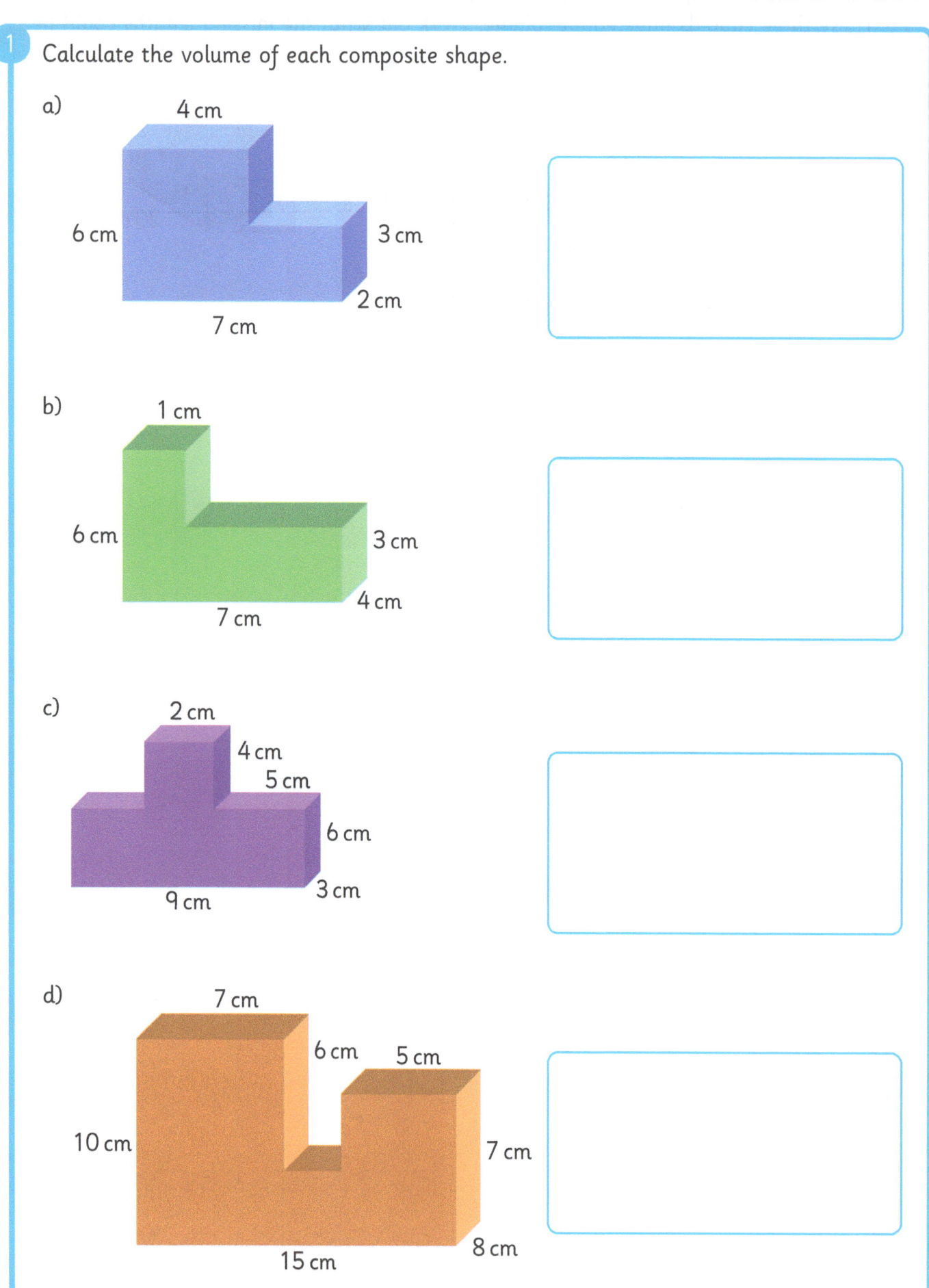

b)

c)

d)

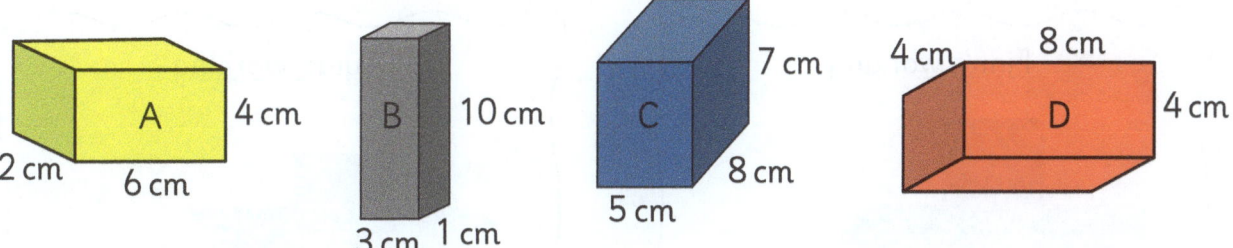

a) Which two of these cuboids can be combined to make a composite shape with a volume of 176 cm³?

b) Three of these shapes are combined to make a composite shape with a volume of 358 cm³. Which shapes are they?

3 Finn and Ayla are working out the volume of a composite shape.

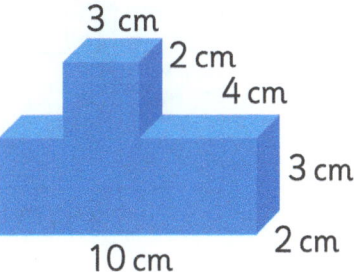

Finn's working is shown on the next page. Ayla says:

I partitioned it into 2 cuboids, not 3.

Show what Ayla's working could have been.

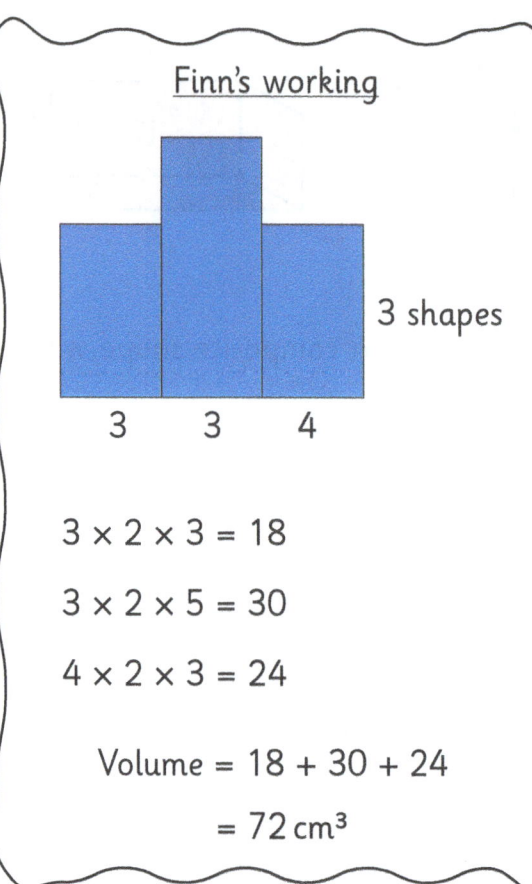

Finn's working

3 shapes

3   3   4

$3 \times 2 \times 3 = 18$

$3 \times 2 \times 5 = 30$

$4 \times 2 \times 3 = 24$

Volume = 18 + 30 + 24

= 72 cm³

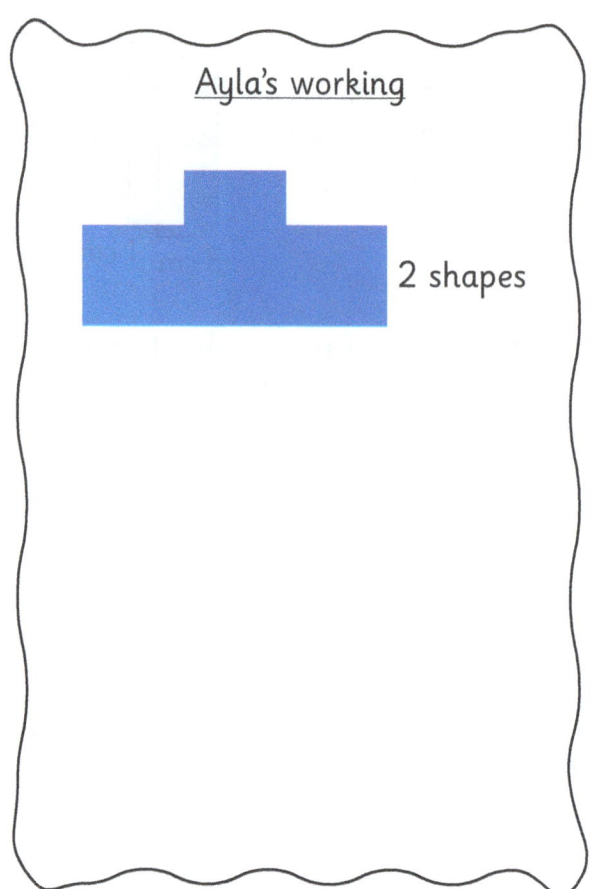

Ayla's working

2 shapes

★ Challenge

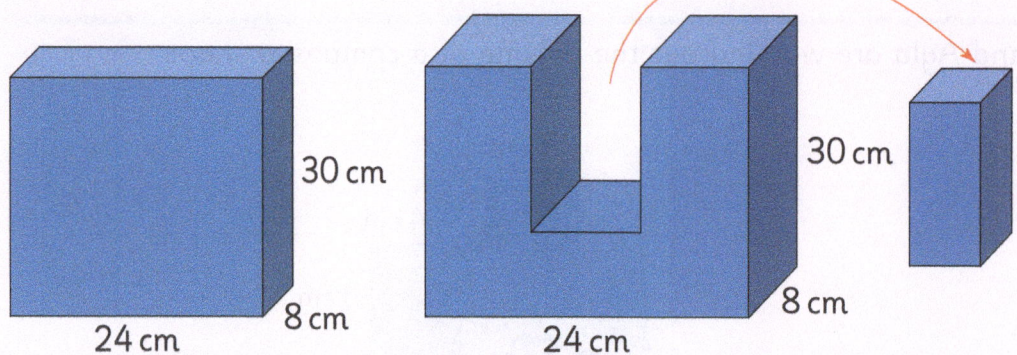

30 cm

8 cm

24 cm

30 cm

8 cm

24 cm

30 cm

A piece of foam measuring 24 cm by 8 cm by 30 cm has a cuboid cut out of it so that it can be used for packing a game in a box. The volume of foam used for packing the game is 5120 cm³. What might the dimensions of the cuboid that was cut out be?

1 Calculate the capacity of each container.

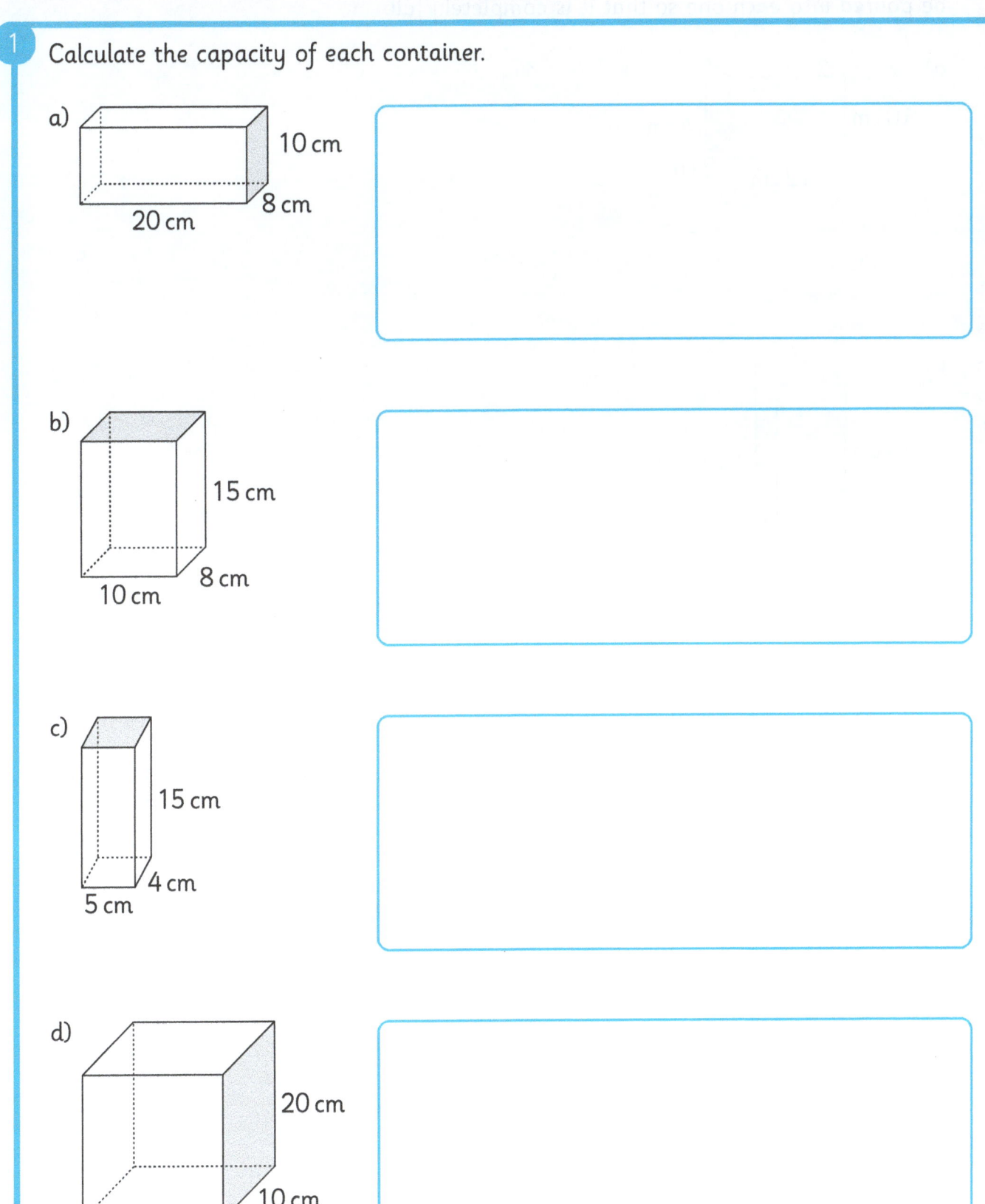

a) 10 cm 8 cm 20 cm

b) 15 cm 8 cm 10 cm

c) 15 cm 4 cm 5 cm

d) 20 cm 10 cm 18 cm

2 Each tank has been partly filled with water. Work out how much water will need to be poured into each one so that it is completely full.

a)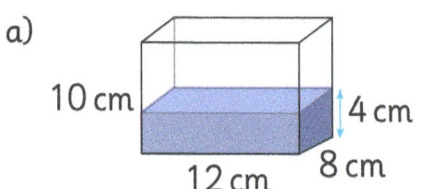

b)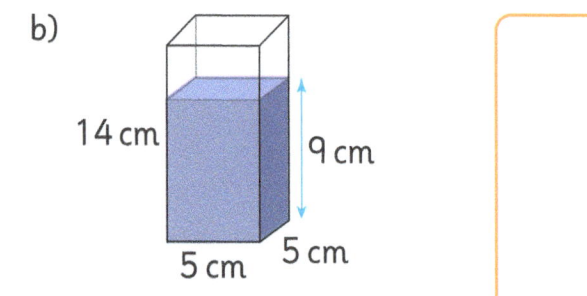

c)

d)

3 Which of these tanks has the greater capacity? Explain your answer.

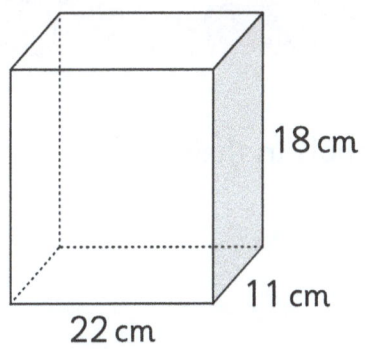

18 cm

11 cm

22 cm

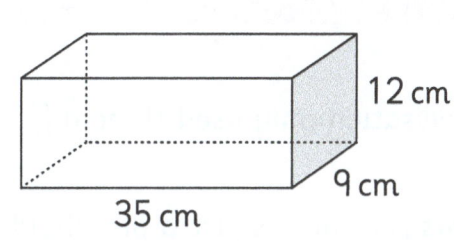

12 cm

9 cm

35 cm

★ **Challenge**

A large fish tank in an aquarium is shown here.

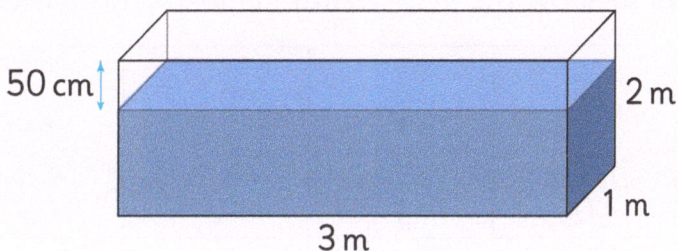

50 cm

2 m

1 m

3 m

a) The water in the tank is filled to 50 cm below the top of the tank. What is the capacity of water in the tank in litres?

b) The number of goldfish in a tank is recommended as 1 per 75 litres. How many goldfish is it safe to put into this tank?

10.1 Mathematics, its impact on the world, past, present and future

The ancient Maya civilisation only used three different symbols in their number system:

⬭ represents zero and is just a placeholder

• represents ones

——— represents a group of five ones

For example:

• • • represents 3 in the Maya number system because it is three ones.

• • •
———
represents 8 in the Maya number system because it is a group of five ones and another three ones.

•
═══
represents 11 in the Maya number system because it is two groups of five ones and another one.

1 Complete this table to show the first twenty numbers in the Maya number system:

⬭	•		• • •
0	1	2	3
	———		
4	5	6	7
• • • ———			• ═══
8	9	10	11
12	13	14	15
			• • • • ═══
16	17	18	19

2 Some students investigate how numbers bigger than 19 were written in the Maya number system. They discover that groups of 20 were used and this was shown using two rows, one above the other.

Number of 20s
Number of 1s and 5s

represents 20   represents 26   represents 30

Write these numbers using the Maya number system.

20   23   25   27

28   30   32   34

★ Challenge

19 is the largest number that can be made using one row:

Mirren and Ciaran are making numbers using two rows:

Mirren: I have got 119. Five twenties on the top row and nineteen ones on the bottom row.

My number is larger. It is 199. I have nine twenties on the top row and nineteen ones on the bottom row. — Ciaran

What do you think the largest number is that can be made using two rows?

11.1 Applying knowledge of multiples, square numbers and triangular numbers to generate number patterns

1 These numbers have fallen out of their envelopes and are mixed up. Each envelope should have exactly four numbers that are part of a sequence of multiples in it.

Multiples of 4   Multiples of 5   Multiples of 7

16   7   8   21   28
20   35   30   12   14   25

a) Write each number in the correct envelope. One has been done for you.

Multiples of 4

40

Multiples of 5

Multiples of 7

b) Write your answers to part a) in order, starting with the smallest, then continue the sequence by writing the next two numbers:

Multiples of 4

Multiples of 5

Multiples of 7

2 Some students are showing the triangular numbers using blocks:

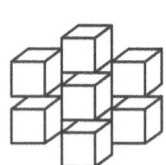
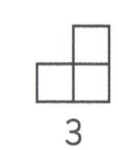
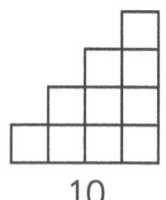

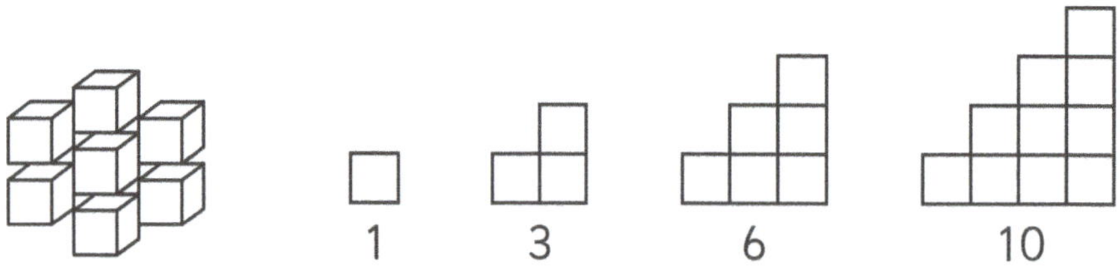

1   3   6   10

a) Draw the next triangular number using blocks.

b) Blair counts the blocks that are left over.

He says:

We have 20 blocks left.

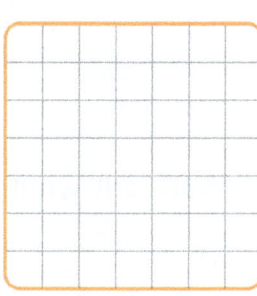

Do they have enough blocks to show the next triangular number?
Explain your answer.

⭐ Challenge

The students read this:

**Every square number can be written as the
sum of two triangular numbers.**

They decide to try this out for some square numbers. Can you help Holly and Jax
with their working?

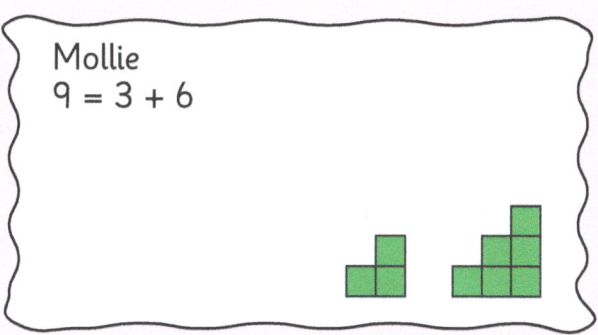

James
25 = 15 + 10

Mollie
9 = 3 + 6

Holly
36

Jax
81

12.1 Solving equations with inequalities

1 These scales balance because the numbers on each side have the same total. This means we can work out what the **?** stands for.

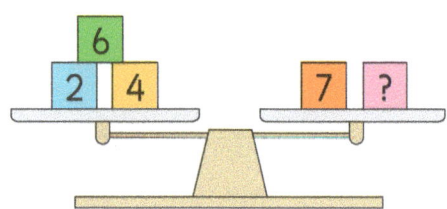

$6 + 2 + 4 = 12$ so, **?** is 5

Work out what the **?** stands for in these balances:

a)

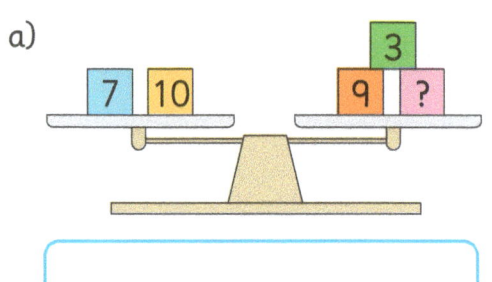

b)

c)

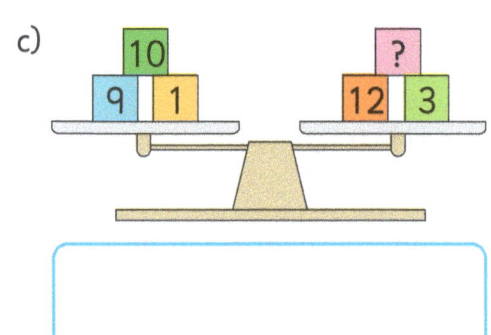

d)

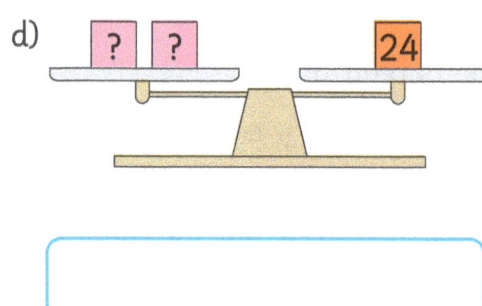

2 Find the missing numbers:

a)

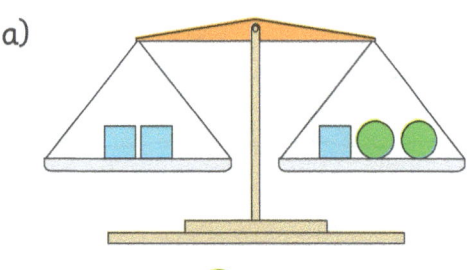

Find ⬤ if ◻ = 12

b)

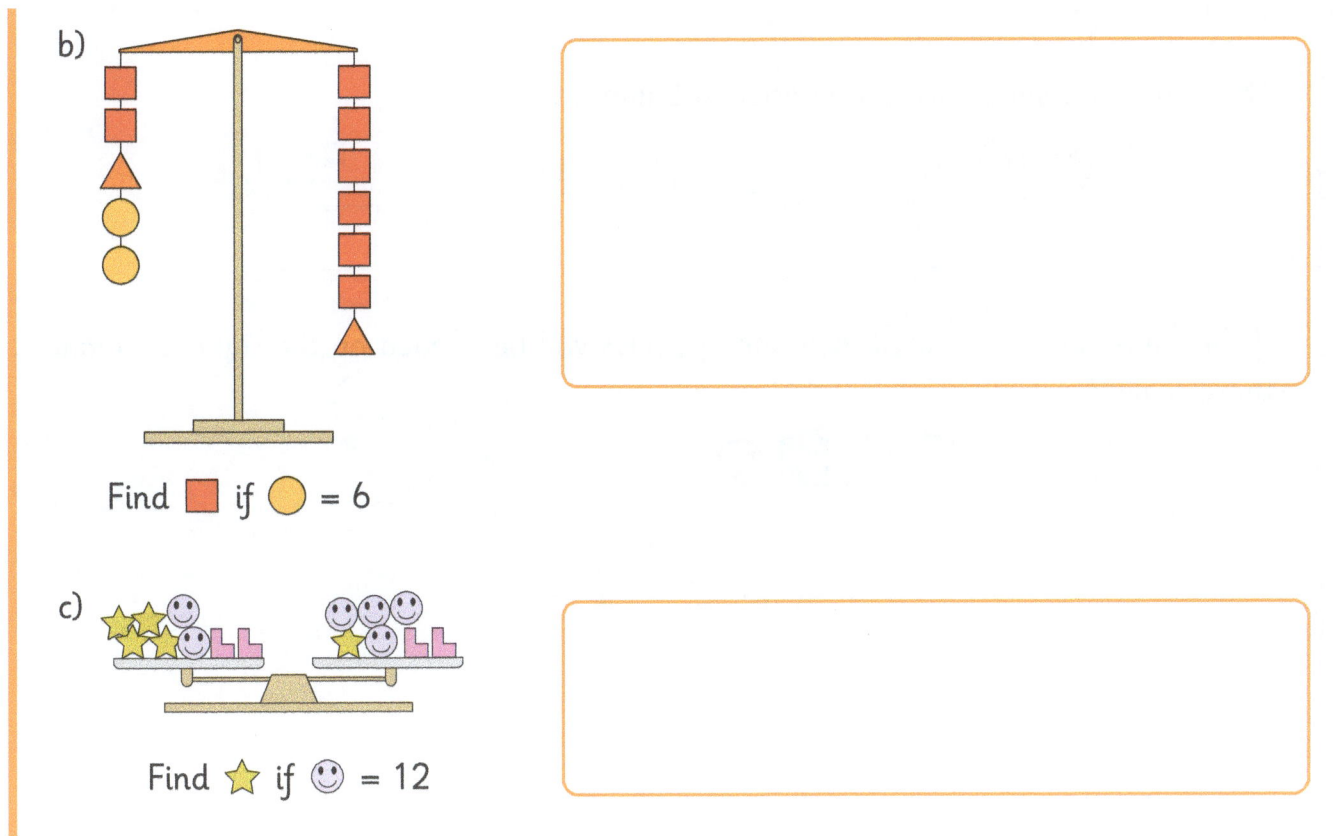

Find ■ if ● = 6

c)

Find ⭐ if 🙂 = 12

3

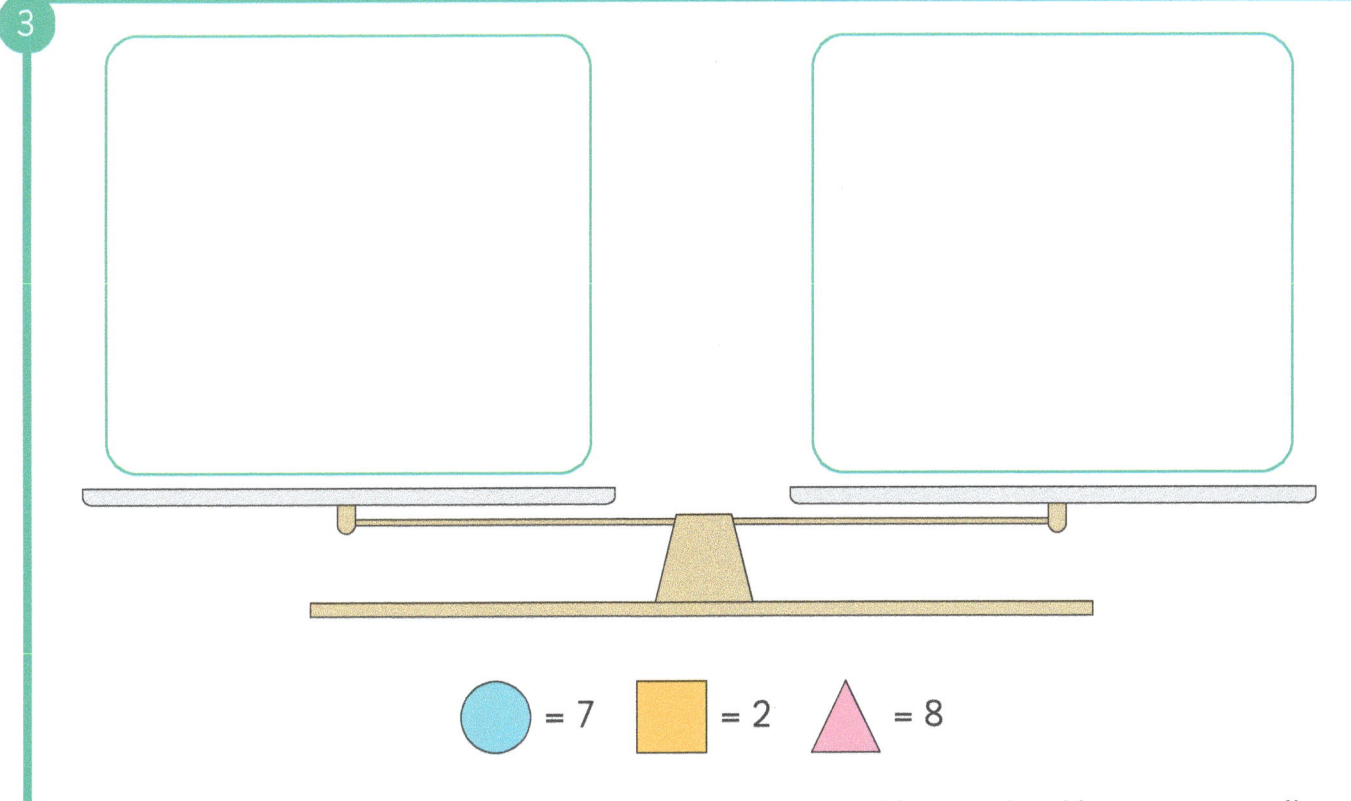

● = 7   ■ = 2   ▲ = 8

Make this scale balance using 4 circles, 6 squares and 10 triangles. You must use all the shapes. Draw your solution in the answer boxes on the scales.

The two scales shown here are perfectly balanced.

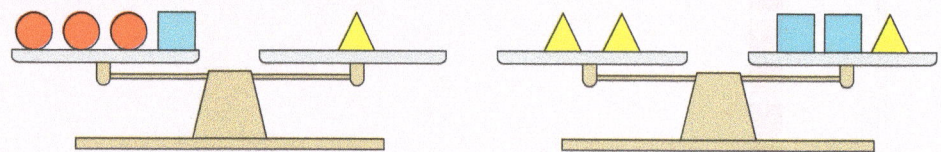

If the same shapes are used, how many circles will be needed on the right to balance these scales?

1 Write the letter in each triangle in the correct column in the table.

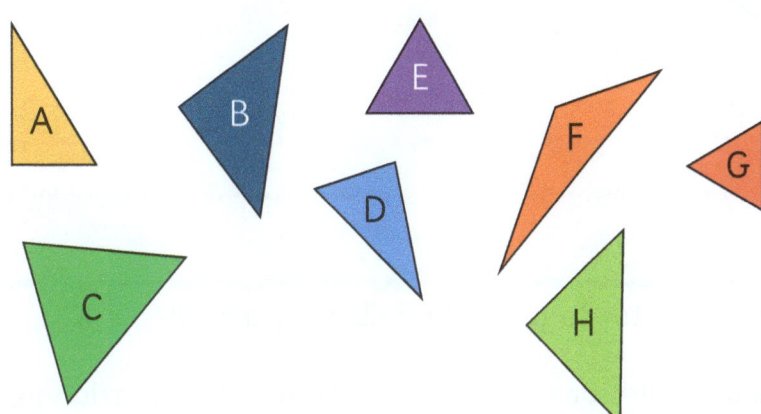

Scalene	Isosceles	Equilateral

2 Use a ruler and a protractor to help you to work out what kind of triangle each of these is.

a)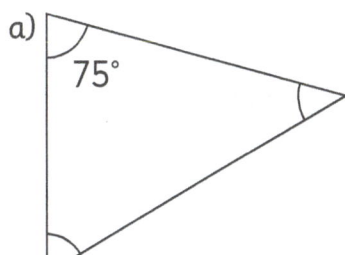
75°

b)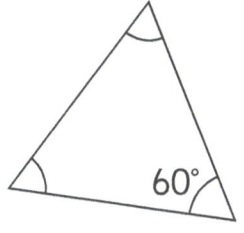
60°

c)
90°

d)

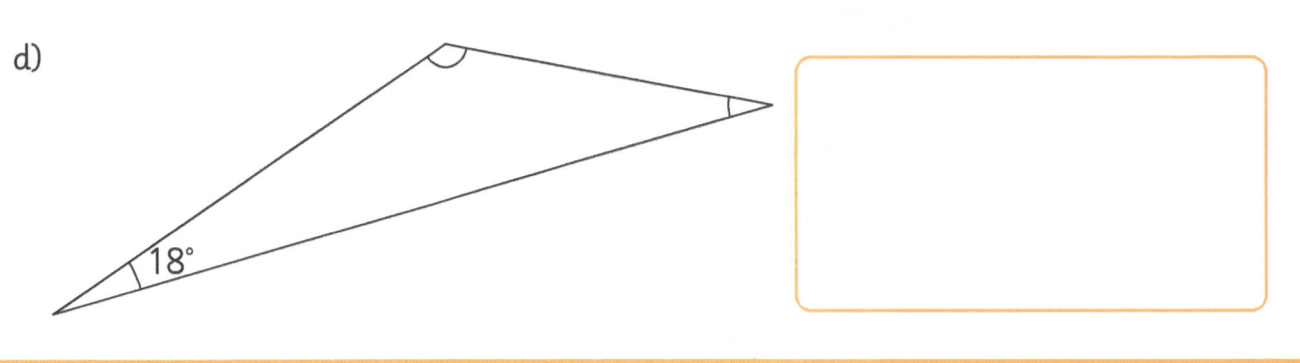

18°

---

3 Use a ruler and protractor to draw the following triangles:

a) A scalene triangle with one 50° angle.

b) An isosceles triangle with a base that is 4 cm long.

c) A scalene triangle with one side of 5 cm and one 110° angle.

d) An equilateral triangle with sides that are longer than 3 cm.

a) Join 3 points on this circle to make an isosceles triangle.

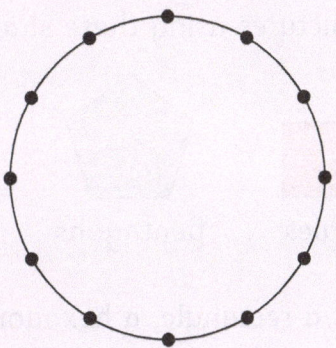

b) Find another two ways to make an isosceles triangle by joining 3 points on the circle.

Do you notice anything?

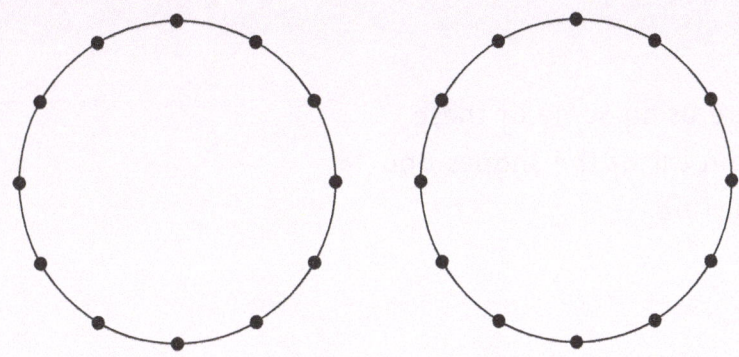

c) Join 3 points on each circle to make:

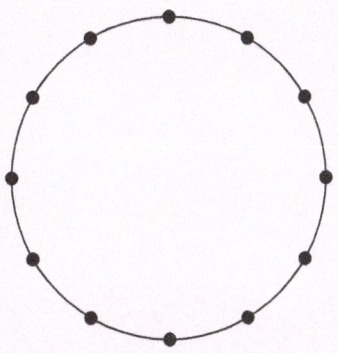
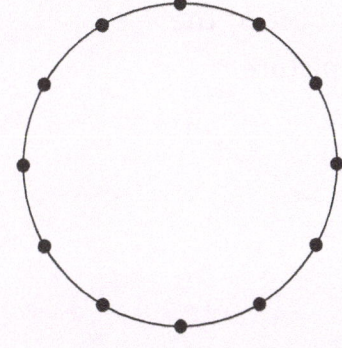
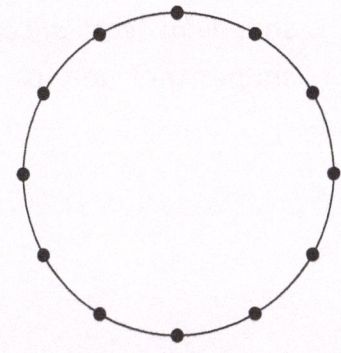

an equilateral triangle   a scalene triangle   a right-angled triangle.

1 Some students are drawing pictures using these shapes:

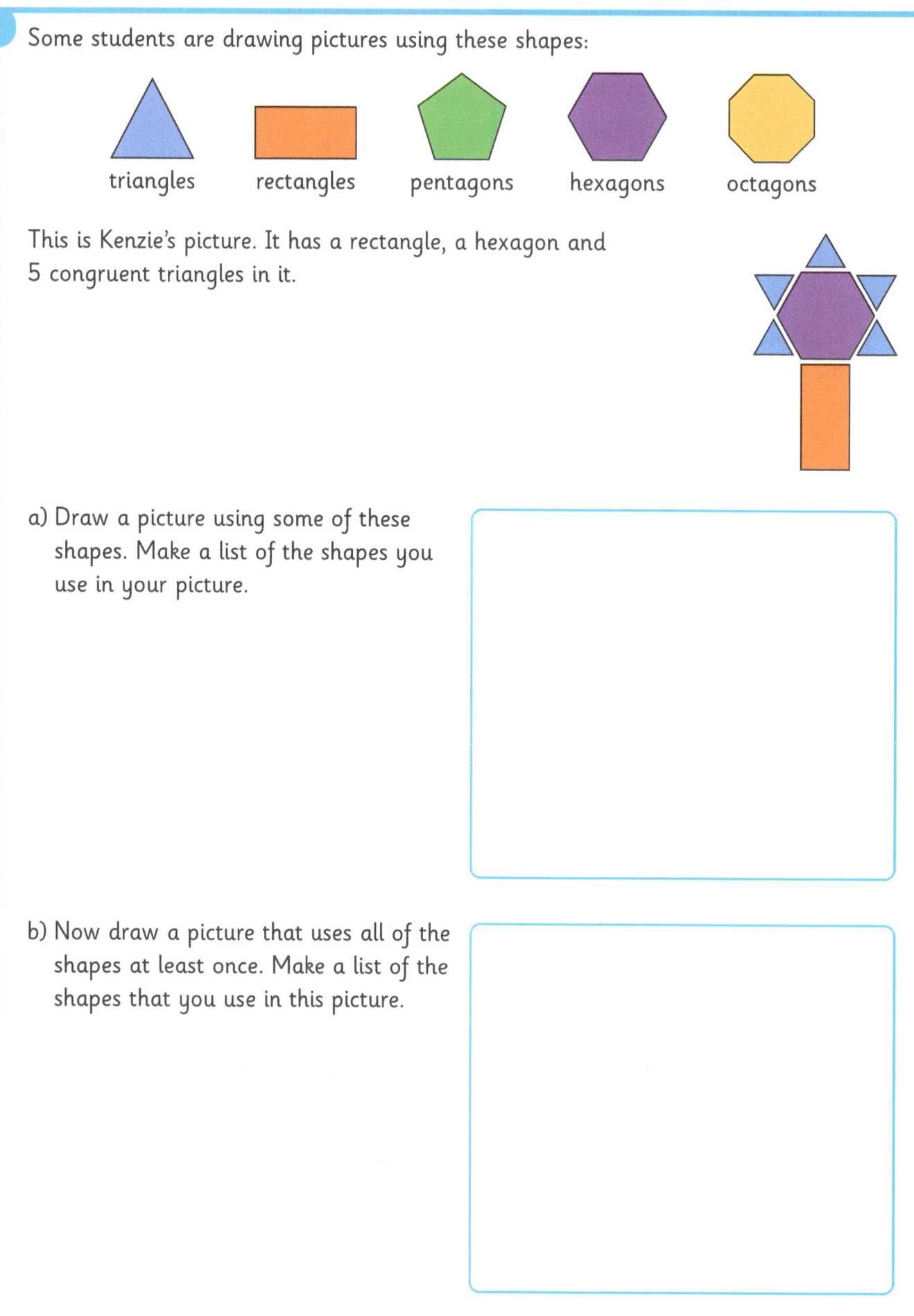

triangles   rectangles   pentagons   hexagons   octagons

This is Kenzie's picture. It has a rectangle, a hexagon and 5 congruent triangles in it.

a) Draw a picture using some of these shapes. Make a list of the shapes you use in your picture.

b) Now draw a picture that uses all of the shapes at least once. Make a list of the shapes that you use in this picture.

2

a) Sketch a shape that has five straight sides of different lengths and five vertices.
Write the name of the shape.

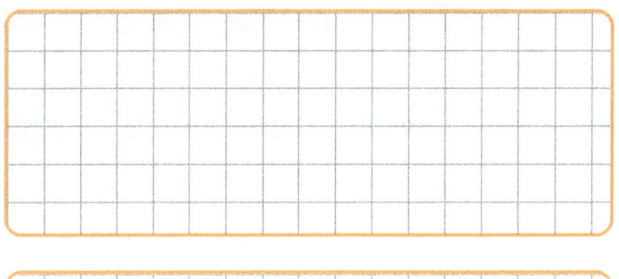

b) Sketch a shape that has four straight sides of equal length and four vertices that are **not** right angles.
Write the name of the shape.

3

a) Draw three more lines to make a square.

b) Draw two more lines to make a scalene triangle.

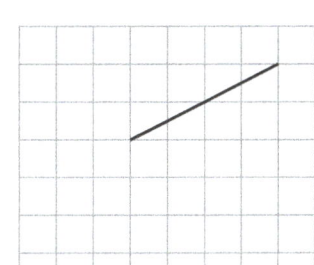

c) Draw two more lines to make a rectangle.

★ Challenge

You can use sticks or straws to help you.

a) This stick pattern has six squares, five small and one large. Without moving any of the other sticks, can you take away three sticks to leave only three squares? Put a cross on each stick on the diagram that needs to be removed.

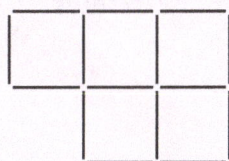

b) This pattern has eight small equilateral triangles. It is made from sixteen sticks. Without moving any of the other sticks, can you take away four sticks to leave four small equilateral triangles? Put a cross on each stick on the diagram that needs to be removed.

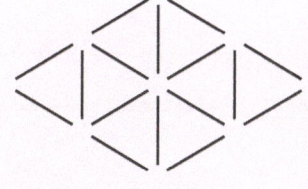

13.3 Making representations of 3D objects

1 Complete each of these partially completed drawings of prisms.

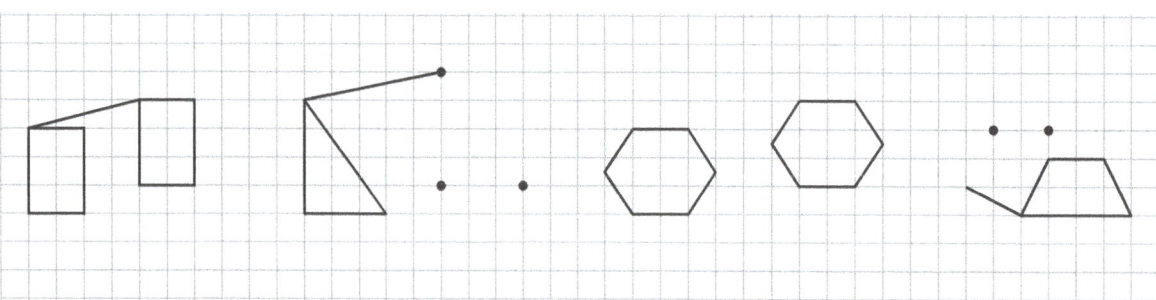

2 a) How many straws and how many blobs of modelling clay are needed to make the skeleton of this cube?

b) Draw a line to match each 3D object to the pile of straws and clay blobs needed to build it. The straws can be cut if required.

3D object	Square based pyramid	Triangular prism	Cuboid	Hexagonal prism
Pile of straws and clay blobs	9 straws 6 blobs	12 straws 8 blobs	8 straws 5 blobs	18 straws 12 blobs

3 Name each 3D object being described here. Sketch each one.

a) This 3D object has six square faces and eight vertices.

Name

b) This 3D object has three rectangular faces, two triangular faces and six vertices.

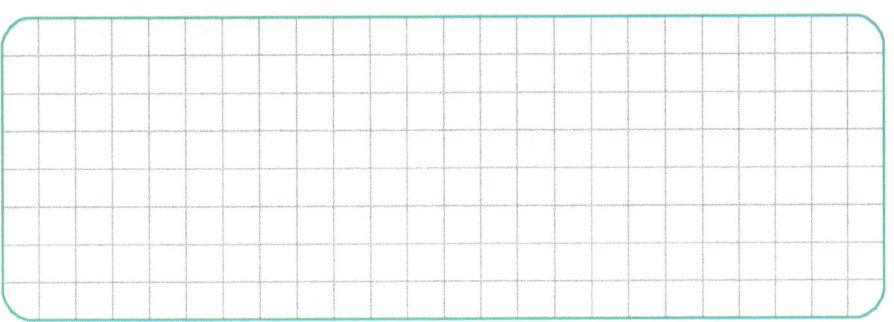

Name

c) This 3D object has four triangular faces, one square face and five vertices.

Name

★ Challenge

An architect uses exactly 1·5 m of wire to make this triangular prism as part of a model.

How long is the prism?

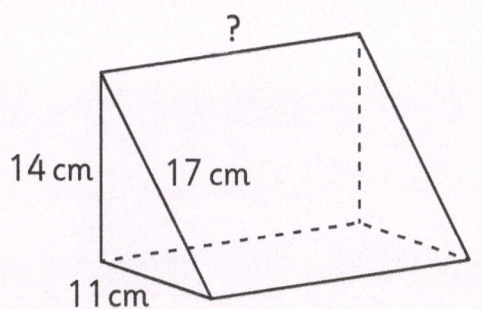

14 cm   17 cm   11 cm   ?

13.4 Nets of prisms

1 Name these 3D objects from their nets:

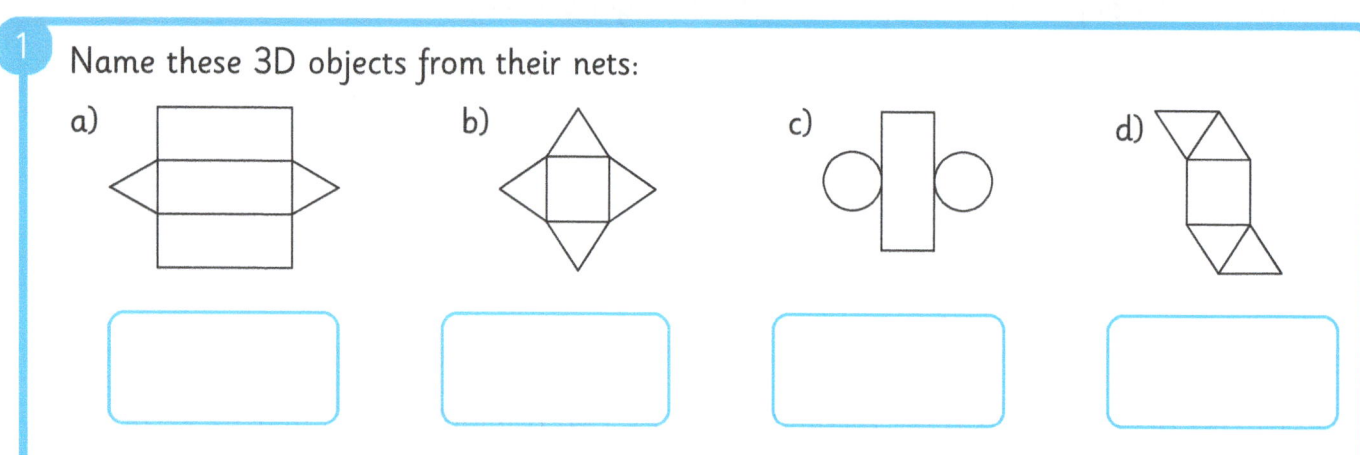

a)

b)

c)

d)

2 Draw an accurate net for each of these 3D objects.

Cuboid

a)
5 cm
2 cm
1 cm

Cube

b)
3 cm

Triangular prism

c)

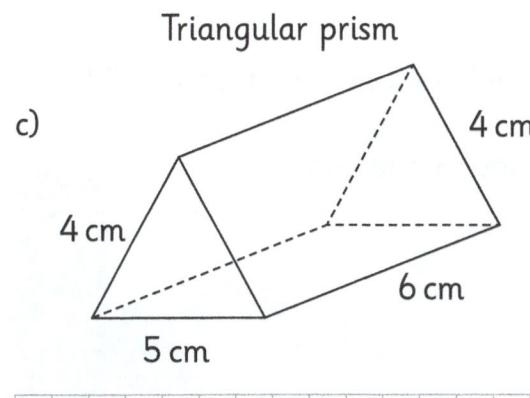

3) This cuboid is drawn on isometric paper. Draw its net on the grid.

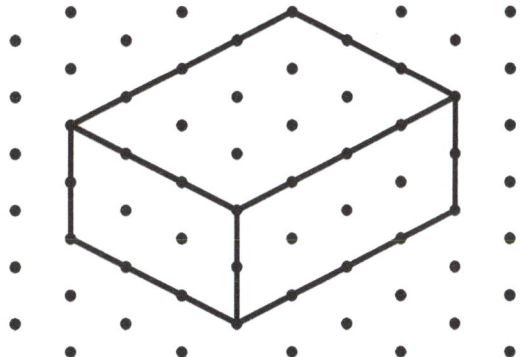

As part of a design challenge some students are making a model of a tower block.

They make nets of a cuboid with a triangular prism on one end.

Lachlan and Miley

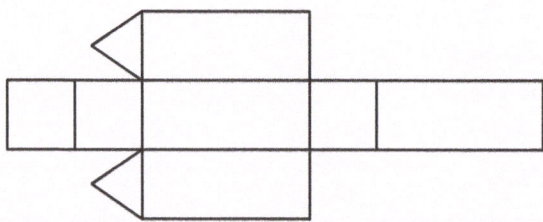

Elliot and Naomi

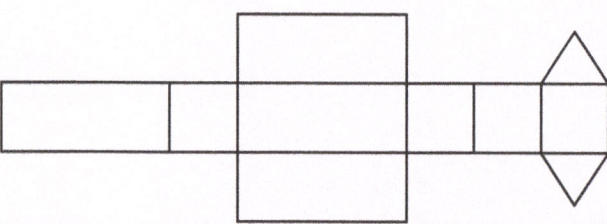

a) When they cut out the nets and try to build the 3D object, only one net works. Which one works? Explain what happens with the other one?

b) Draw your own net for this object.

1 Draw each triangle accurately using a ruler and a protractor then measure the missing angle in each one. (The diagrams are not drawn to scale.)

a)

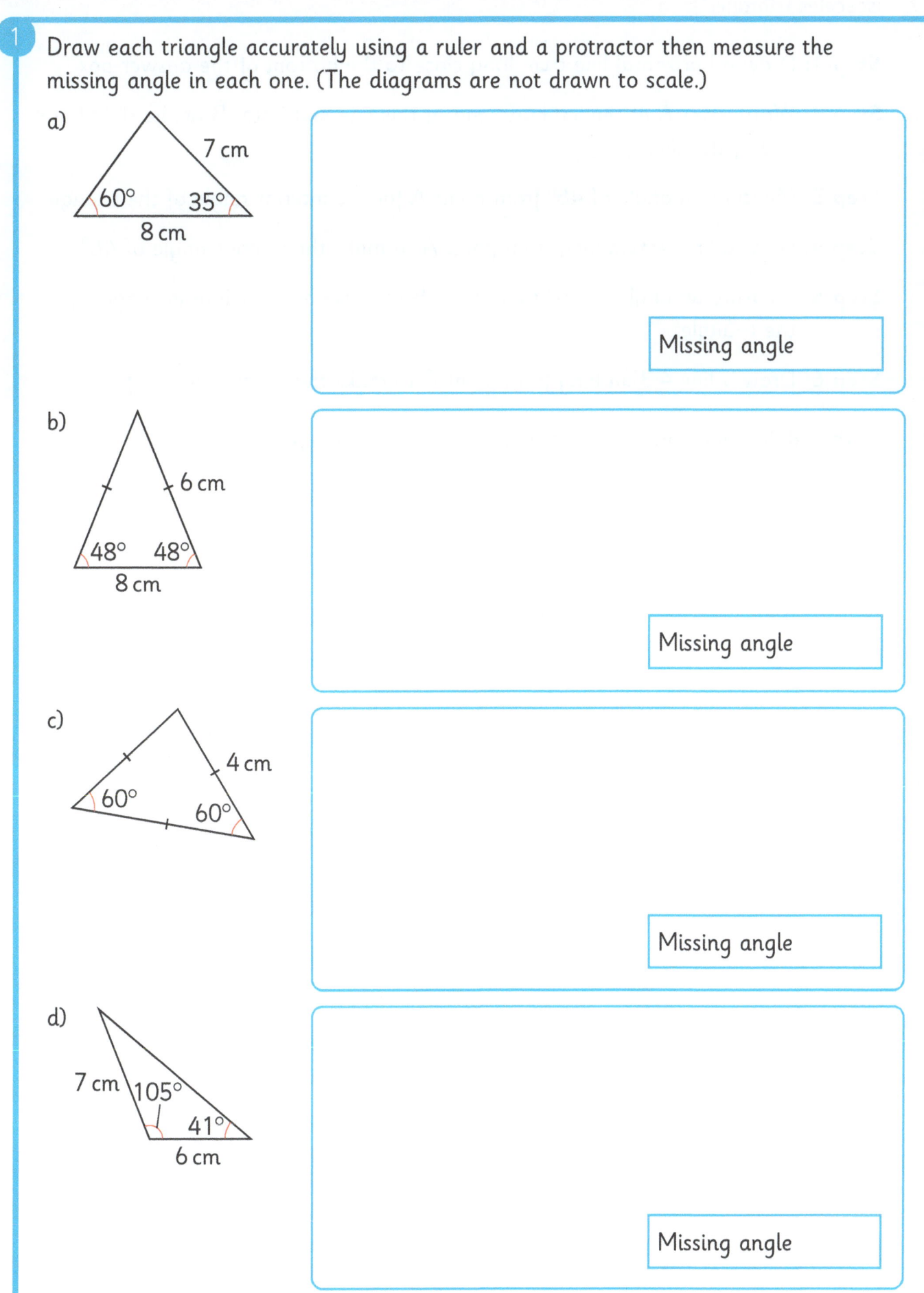

Missing angle

b)

Missing angle

c)

Missing angle

d)

Missing angle

2 Some students are following these instructions to make an accurate drawing of an isosceles triangle:

Step 1: Draw a horizontal line 6 cm long close to the bottom of the answer box.

Step 2: Write letter A at the left-hand end of this line and letter B at the right-hand end of this line.

Step 3: Measure an angle of 46° from point A for the interior angle of the triangle.

Step 4: Draw a line 4·3 cm long from point A to make the interior angle of 46°.

Step 5: Measure an angle of 46° from point B to make another interior angle of the triangle.

Step 6: Draw a line 4·3 cm long from point B to make the interior angle of 46°.

Faith and Jasmine both went wrong. Here are their attempts:

Faith   Jasmine

a) Follow the steps in the instructions to draw the triangle correctly using a ruler and a protractor.

b) Use a protractor to measure the size
   of the third angle in your triangle.

c) Describe what Faith and Jasmine did wrong.

Accurately draw each of these triangles in the box below (continue on a separate piece of paper if you need to):

a) two different isosceles triangles with an angle of 40°

b) a scalene triangle with sides measuring 5 cm, 5·5 cm and 6 cm

c) a right-angled triangle with sides measuring 15 mm, 36 mm and 39 mm

d) an equilateral triangle with sides measuring 4 cm.

1 Find the reflex angles that are **inside** this shape. Mark them with an arc on the diagram.

2 a) Measure the acute angle and the reflex angle here.

b) Measure the obtuse angle and the reflex angle here.

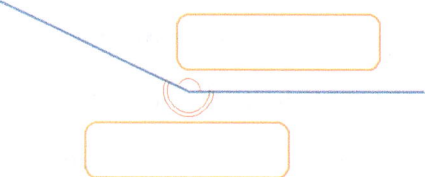

Measure these reflex angles.

c)

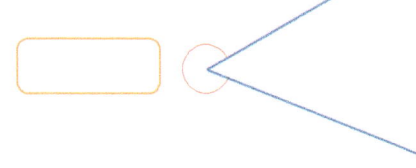

d)

e)

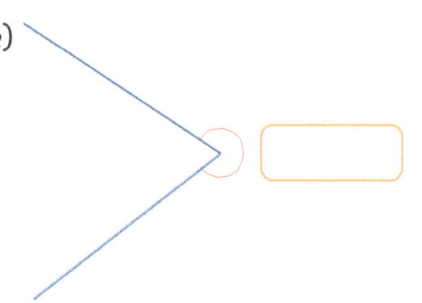

f)

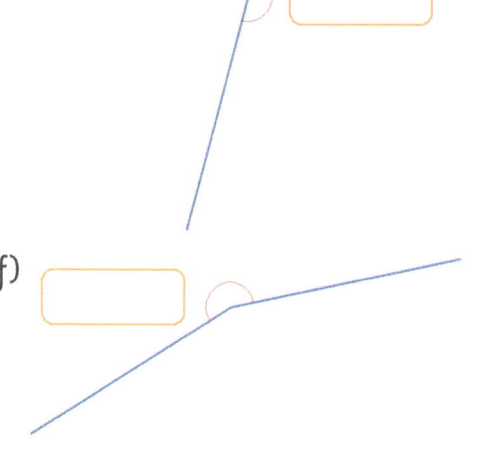

3 Draw the following reflex angles.

a) 250°

b) 305°

★ Challenge

a) Draw an example of a quadrilateral (a four-sided shape) that has a reflex angle as one of its **interior** angles.

b) Draw an example of a pentagon (a five-sided shape) with reflex angles for two of its **interior** angles.

c) Draw a shape with several **interior reflex** angles. Can you find out and write the name of your shape?

14.3 Finding missing angles

1 Calculate the missing angles:

100° A 110°

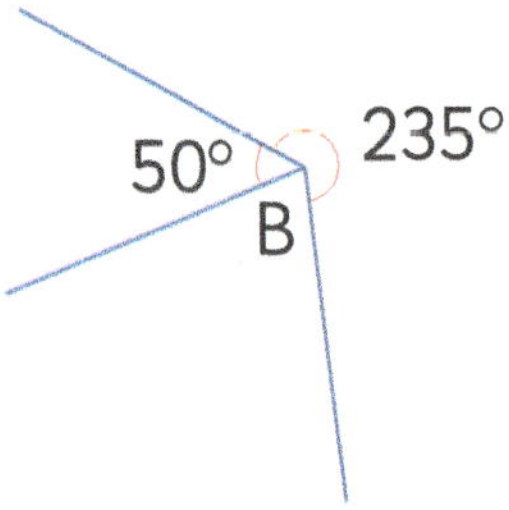

50° B 235°

C 106° 97° 101°

130° D E 130°

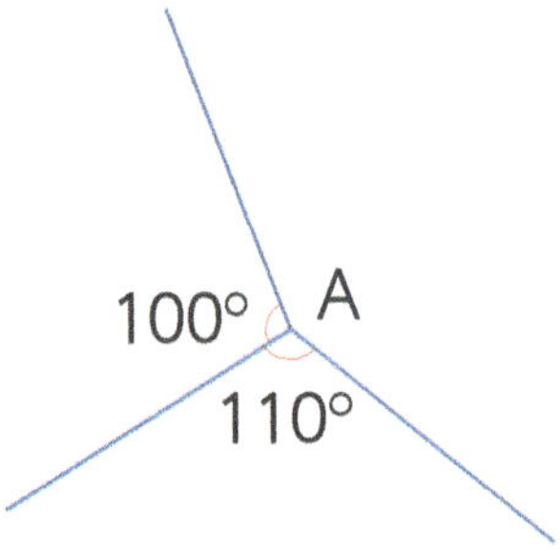

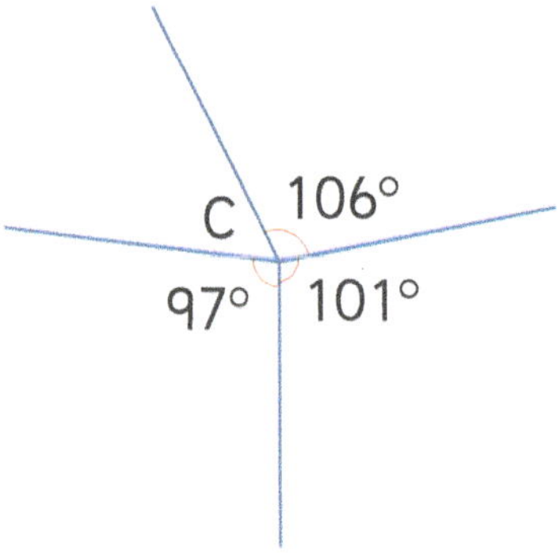

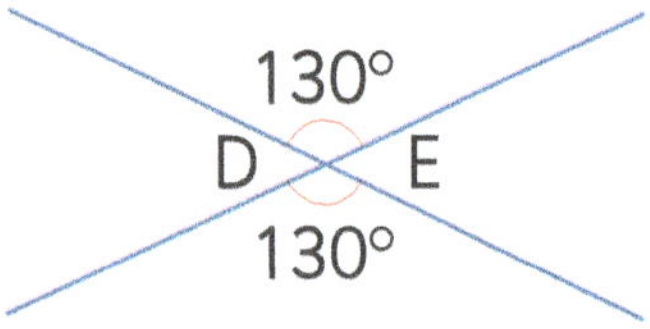

2 Calculate the missing angles:

F 40°

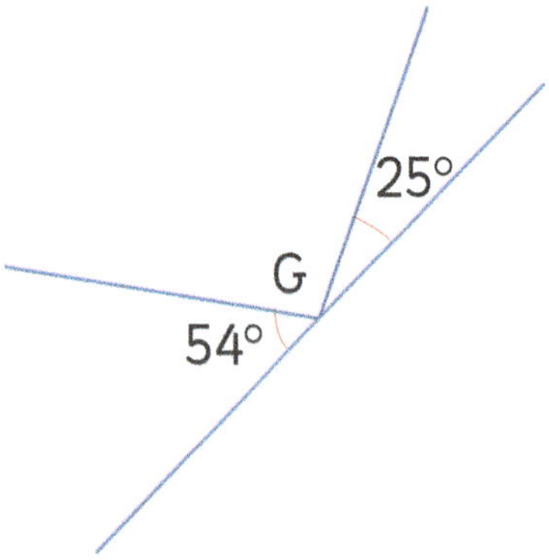

25° G 54°

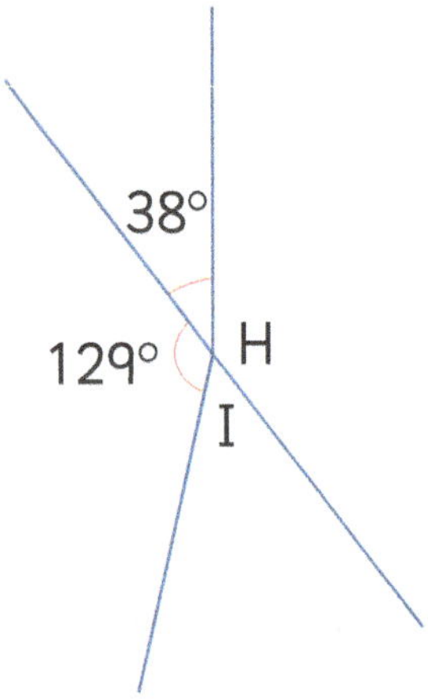

38° 129° H I

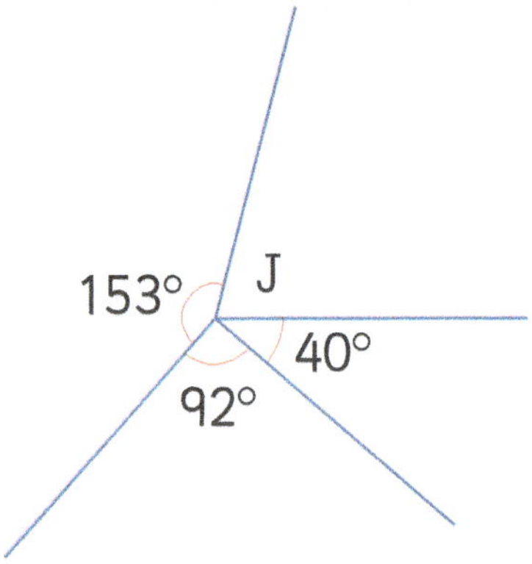

153° J 40° 92°

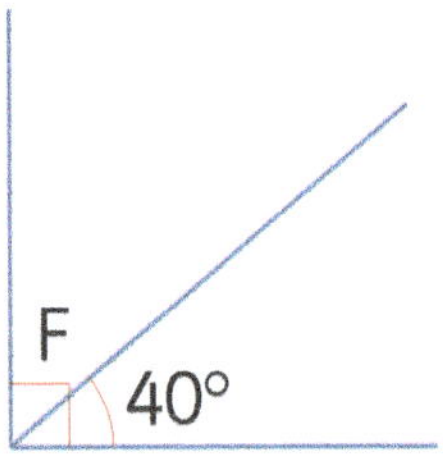

3 Here are some angle cards.

A 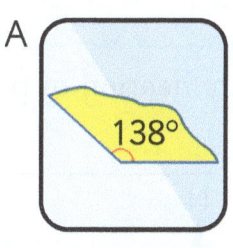 138°

B 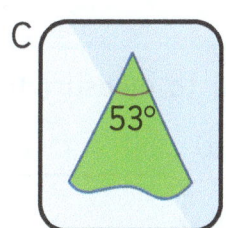 37°

C 53°

D 60°

E 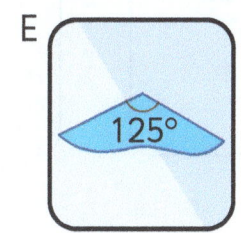 125°

F 67°

a) Which two cards fit together exactly to make a right angle?

b) Which three cards fit together exactly on a straight line?

c) Which four cards fit together exactly around a point?

⭐ **Challenge**

Use the digit cards below each diagram to fill in the missing numbers. You can only use each digit card once.

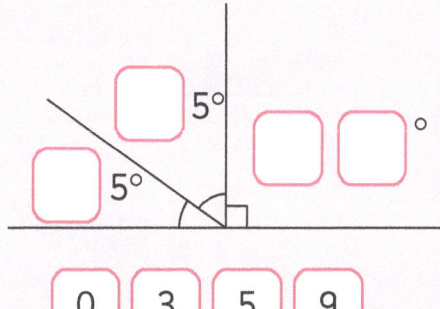

☐5° ☐5° ☐☐°

0 3 5 9

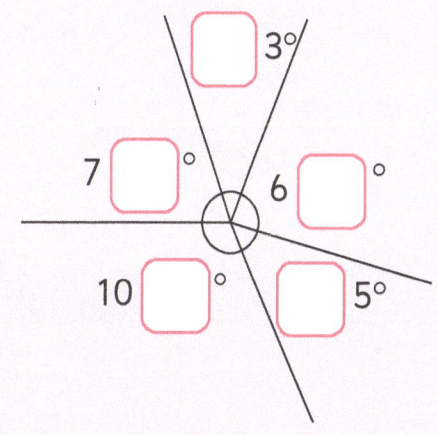

☐3° 7☐° 6☐° 10☐° ☐5°

1 2 5 6 9

1 Use a protractor to measure then record the 3-figure bearing of point B from point A.

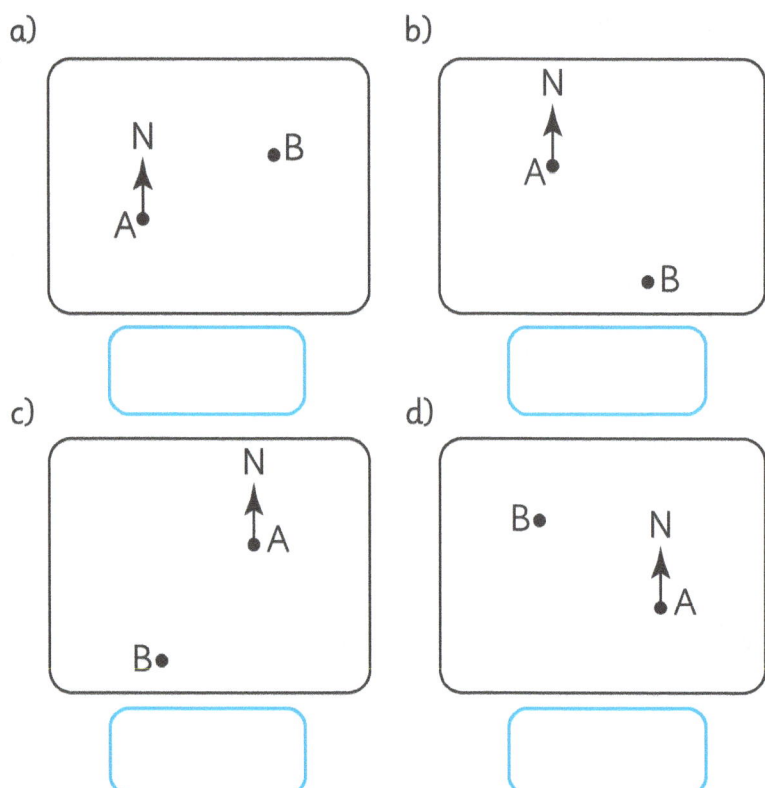

a)

b)

c)

d)

2 Use a protractor to measure then record the 3-figure bearing of each boat from the lighthouse.

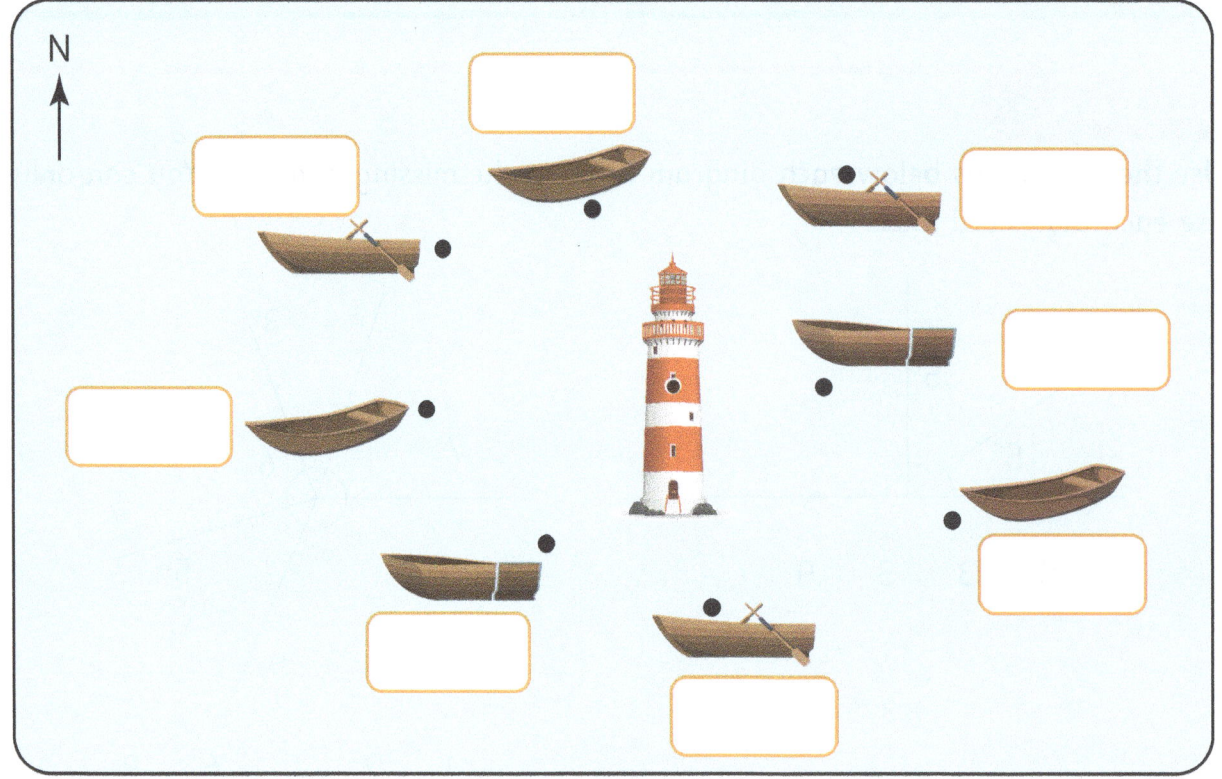

This shows an air traffic control radar screen at an airport. The distance between each circle represents 10 km.

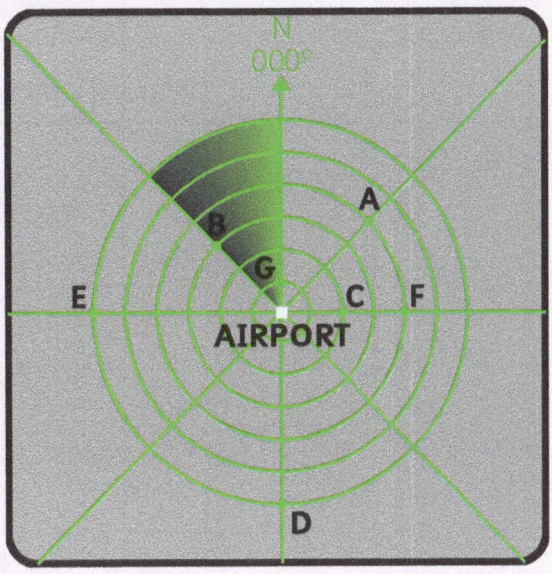

a) Aeroplane A is 40 km from the airport on a bearing of 045°. How far is aeroplane B from the airport and what is its 3-figure bearing?

b) Aeroplane A and aeroplane F are both 40 km from the airport. Identify another pair of aeroplanes that are the same distance away from the airport and give the 3-figure bearing for each one.

c) The aeroplane that is closest to the airport is preparing to land. Which aeroplane is this and what is its bearing?

d) Aeroplane H is 50 km away from the airport and is approaching from the Southwest. Mark aeroplane H on the diagram and write down its 3-figure bearing.

1 This map shows a holiday island with some of the attractions marked.

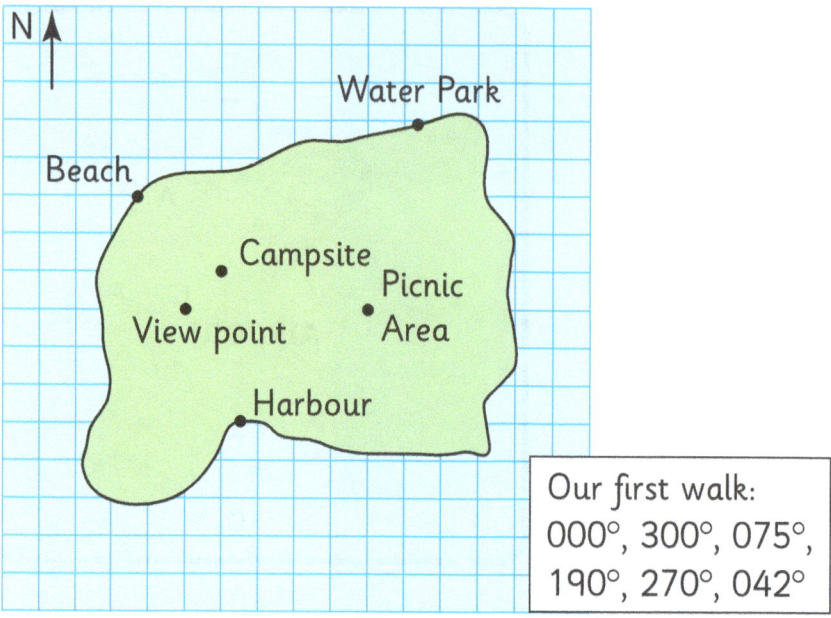

Our first walk:
000°, 300°, 075°,
190°, 270°, 042°

Harvey and his family arrive at the **harbour** by boat. Harvey keeps a note of the 3-figure bearings of their first walk around the island so that he can draw the walk on the map.

Show the family's walk on the map and list the attractions they visited in order.

Harbour

2 Some students have written a code for a small programmable robot to move around on a page.

Follow these instructions and plot the journey using a ruler and protractor.

3 cm on a bearing of 115°, then 7 cm on a bearing of 265°, then 6 cm on a bearing of 025°, then 4 cm on a bearing of 094°, then 5 cm on a bearing of 126°, then 5·5 cm on a bearing of 260°.

• Start

⭐ **Challenge**

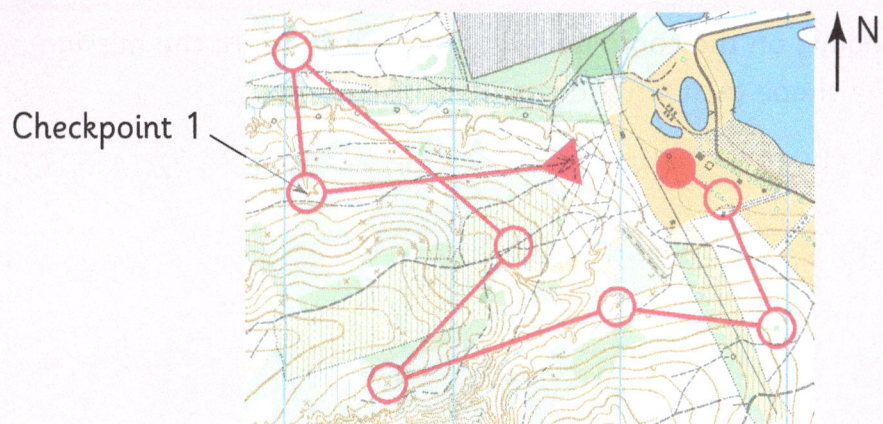

Checkpoint 1

This is a map of an orienteering course. It begins at the ◀ and ends at the ●.
If 1 cm on the map represents 1 km on the ground, complete the table then describe
the journey around the course using three figure bearings and distances in km.

Section of the course	Distance	3-figure bearing
Start to checkpoint 1		

1 Calculate the missing coordinates for each shape.

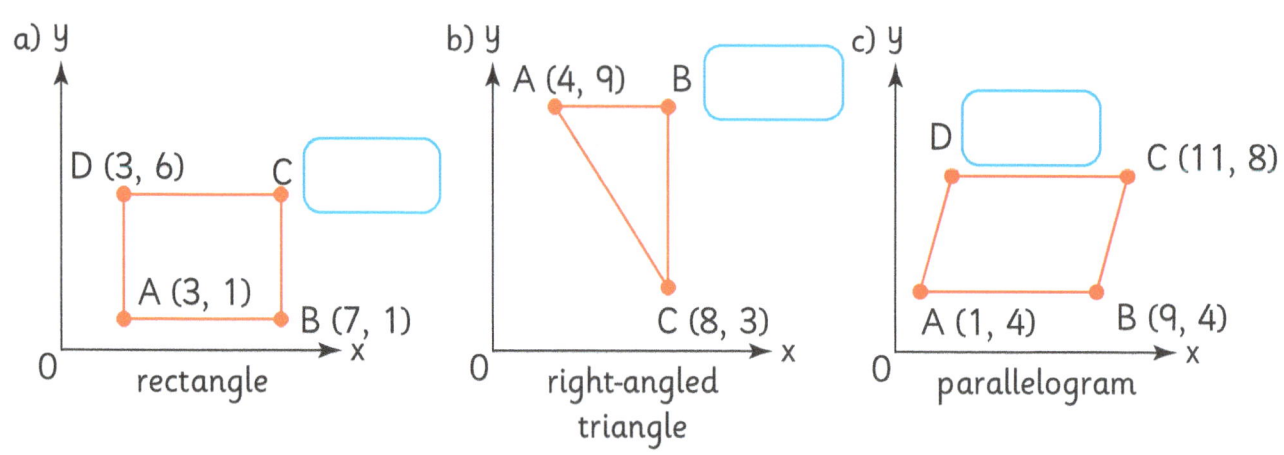

a) rectangle

b) right-angled triangle

c) parallelogram

2 Plot these points on the coordinate diagram given. Mark the missing point to complete the shape then write down its coordinates.

a) Rectangle (1, 2) (5, 2) (5, 5)

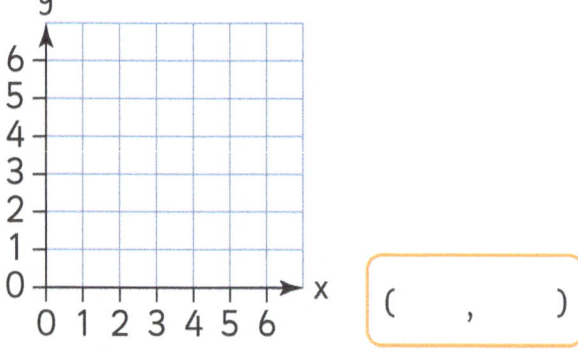

(,)

b) Rhombus (2, 0) (4, 3) (2, 6)

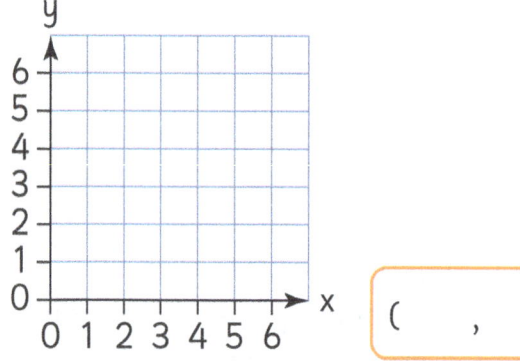

(,)

c) Square (1, 3) (4, 1) (6, 4)

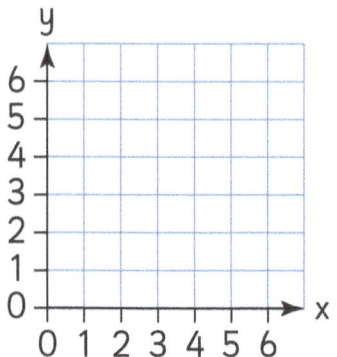

(,)

d) Isosceles triangle (3, 0) (6, 5)

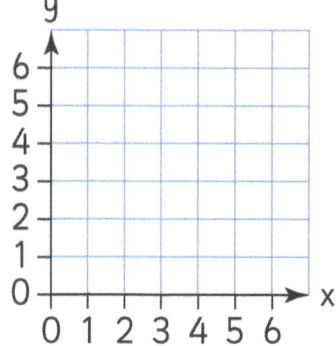

(,)

3 a) The coordinates for two vertices of a kite are shown here. What could the coordinates of the other vertices be?

y

• (4, 9)   • (12, 9)

0 ——————→ x

What else might they be?

b) The coordinates for two vertices of a square are shown here. What could the coordinates of the other vertices be?

y

• (5, 6)

• (5, 2)

0 ——————→ x

What else might they be?

★ Challenge

Three vertices of a hexagon are shown here. Add another three vertices to the diagram to complete a hexagon. Write down the coordinates of each vertex.

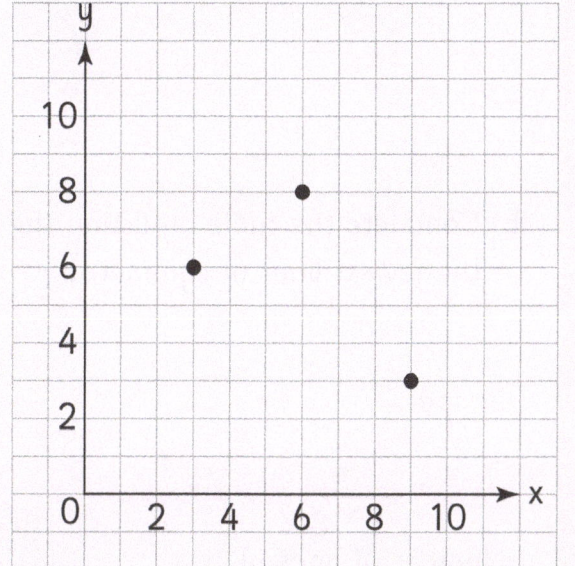

14.7 Symmetry 1

1 Draw all the lines of symmetry onto these shapes.

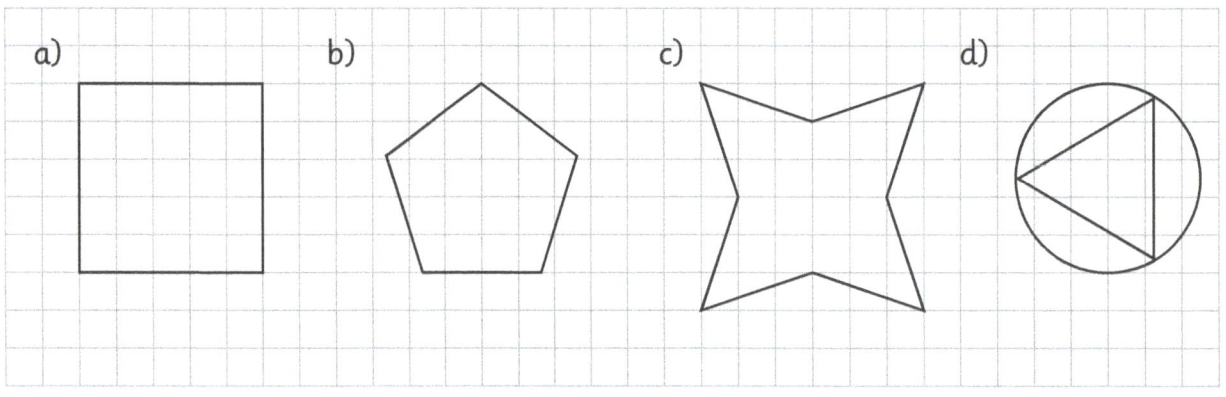

a) b) c) d)

2 a) Work out how many lines of symmetry each shape has and draw them in.

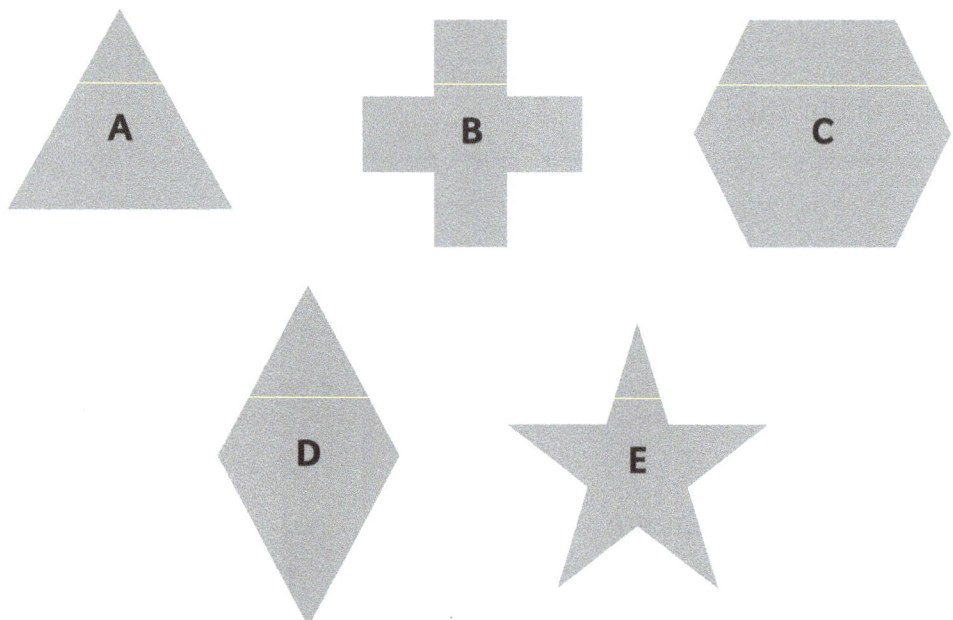

A B C

D E

b) Complete the table by listing the shapes in order, starting with the one that has the fewest lines of symmetry.

Shape					
Number of lines of symmetry					

3 This diagram shows a square drawn inside a regular octagon.

a) Draw all the lines of symmetry on the shape.

b) How do you know that you have found **all** the lines of symmetry?

⭐ **Challenge**

Some students are researching flags. They find out that the flag of Switzerland is a square that has four lines of symmetry.

Design a flag with exactly four lines of symmetry. Use at least two different colours in your flag design.

1 Complete the shapes by reflecting them in the lines of symmetry. Make sure the colours are symmetrical too.

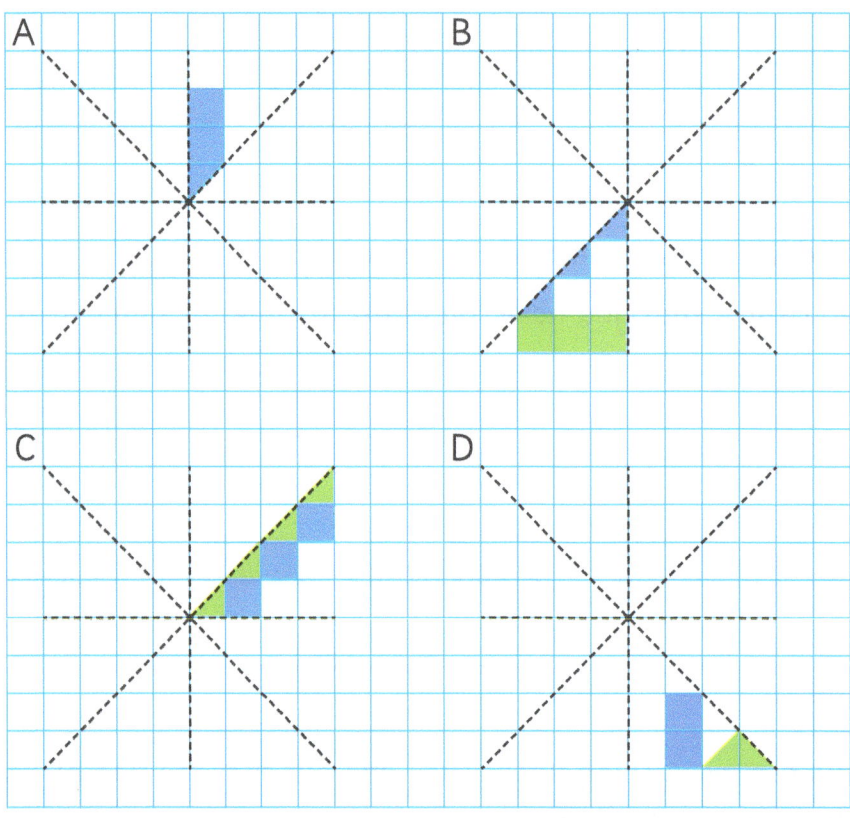

2 Complete this design to give it four lines of symmetry.

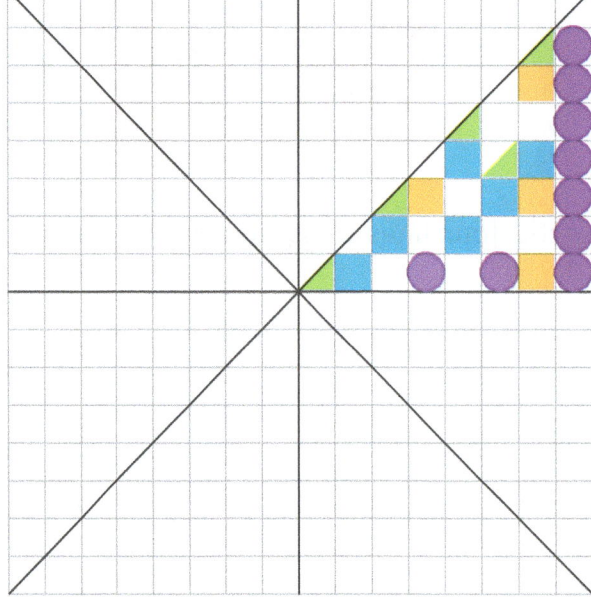

3 Kayleigh says:

> This pattern should have four lines of symmetry, but it doesn't look right.

Alhady says:

> You are correct. I can see some errors.

a) How many errors can you see in the diagram?

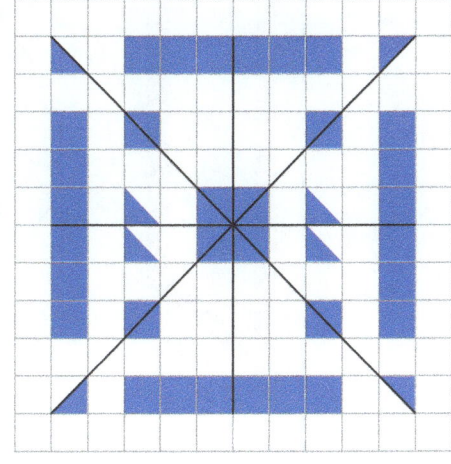

b) Add some shading to the diagram so that it has exactly four lines of symmetry.

⭐ **Challenge**

Use coloured shading to create a symmetrical pattern on this hexagonal design.
How many lines of symmetry can you include in your pattern?

1. This map of a country park is drawn using a scale of 1 cm : 500 m.

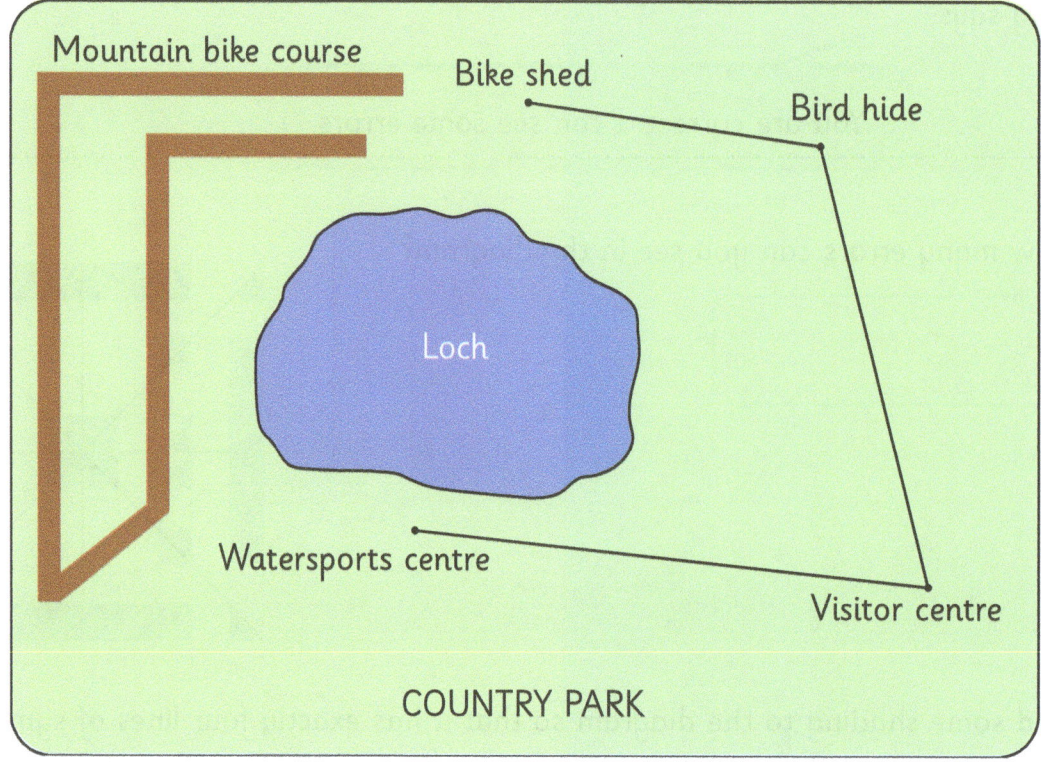

a) What is the actual distance between the visitor centre and the water sports centre?

b) Maria and Cooper visit the country park with their bikes. They cycle from the visitor centre to the bird hide then to the bike shed. How far did they cycle altogether?

c) Maria and Cooper decide to go around the mountain bike course. Work out the length of the course.

2 Some students are drawing a map for the festival. They measure the distances with a trundle wheel and the directions with a compass.

The face painting is 150 m north of the park entrance, the refreshment stall is 200 m north-west of the park entrance, the archery is 125 m south-east of the refreshment stall and the balloon modelling is 175 m south-west of the face painting.

N ↑

X
Park entrance

a) Using a scale of 1 cm : 25 m, and a protractor, mark where the face painting, refreshment stall, archery and balloon modelling areas are accurately on the map.

b) A first aid ✚ area needs to be positioned at least 75 m away from the park entrance and no more than 50 m away from the archery. Add the ✚ symbol to a suitable place for the first aid area on the map.

★ **Challenge**

Some students find an old road atlas in a school cupboard.

They measure the width of each page to be 18 cm and work out from the maps that this represents 63 km.

Find the scale of the maps in the atlas.

GREAT BRITAIN

Easy to read Large scale

15.1 Sampling

1 Write a word or phrase from this list to complete each sentence.

sample	population	big question	data

a) [   ] is a word we use to describe information.

b) In data handling, a [   ] is the entire group information is

being gathered from.

c) A [   ] only gathers data from some of the people who can take part.

d) A [   ] highlights what the researcher wants to find out about.

2 Decide whether each of these surveys gathers data from a whole population or from a sample. Explain your thinking. One has been done for you.

a) A drama group for children aged between 8 and 15 years old asks everyone in the group who is aged 8, 9 and 10 what their favourite musical film is.

Whole population / ⟨Sample⟩	This is a sample because it only involves some of the children. It misses out the older ones.

b) All of the players in a rugby union team are asked to look at three team strip designs and choose their favourite.

Whole population / Sample	

c) Everyone who visits a leisure centre on a Monday afternoon is asked if they would like the café to sell fresh fruit.

Whole population / Sample	

3 Complete this table to show how data could be gathered from a whole population or from a sample to answer a big question. One has been done for you.

Big question	Whole population	Sample
What reading genre do the members of a book club prefer?	All the members of the book club	Every fifth name on an alphabetical list of book club members
How do students in a school travel to school each day?		
Which game do members of a virtual gaming club like best?		
Which breed of dog is most popular in a dog agility club?		

⭐ **Challenge**

A cinema manager wants to find out what genre of film is most popular with adults living in the local area. She asks her staff to speak to people as they leave the cinema, having watched a popular family-friendly animated movie in the school holidays.

a) Do you think the cinema manager will gather useful data here? Explain your thinking.

b) Suggest two alternative ways that the cinema manager might gather data.

1.

2.

1 Say whether each of these graphs is accurate or misleading and explain why.

a) Number of tickets sold

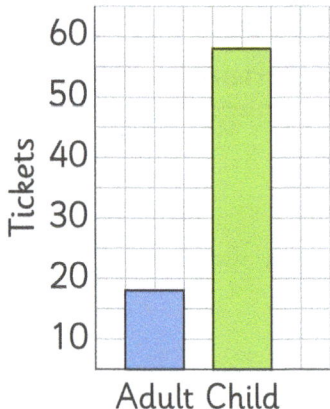

Accurate or misleading?

b) Rainfall over 5 days

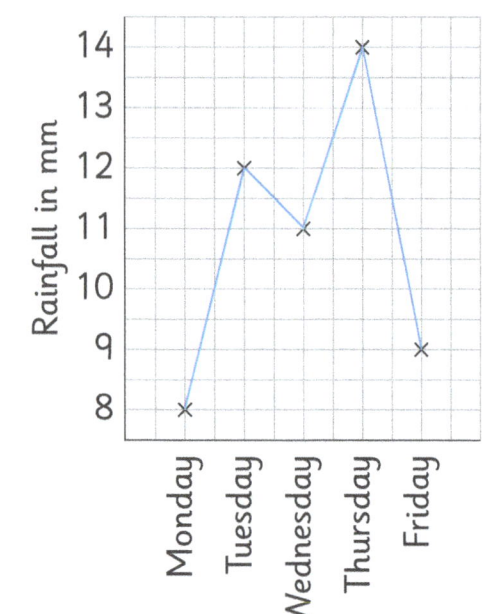

Accurate or misleading?

c) Favourite dance style

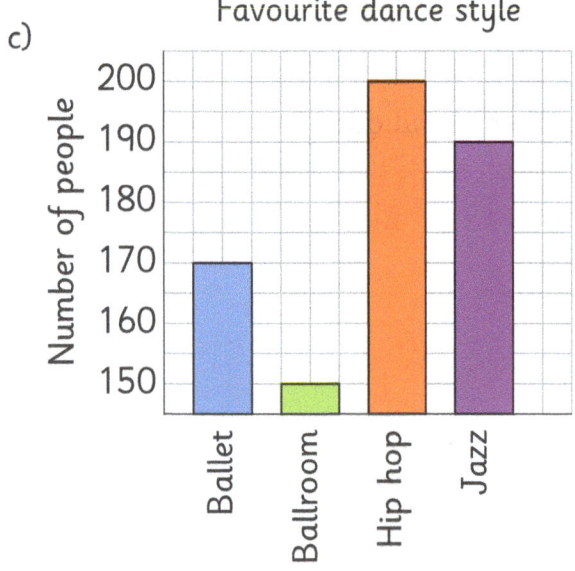

Accurate or misleading?

d) Money raised this year

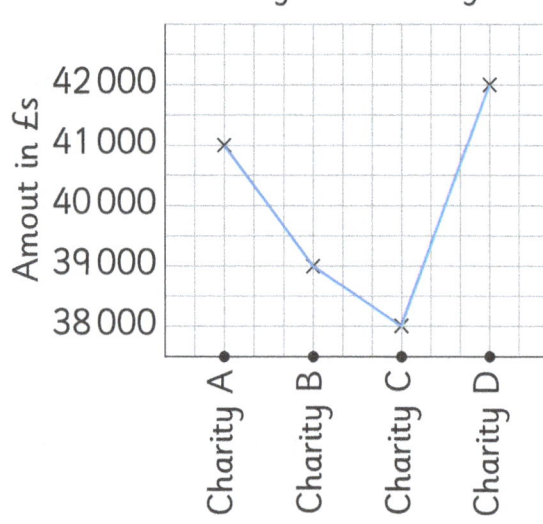

Accurate or misleading?

★ **Challenge**

An advertising company is designing an advert for a new type of compost.
A designer presents two different possibilities, each including a graph based on data provided about the average height of sunflowers planted in the compost.

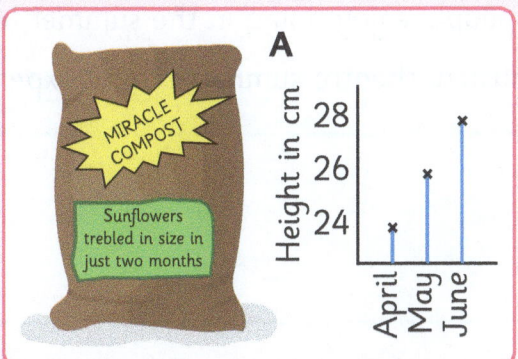

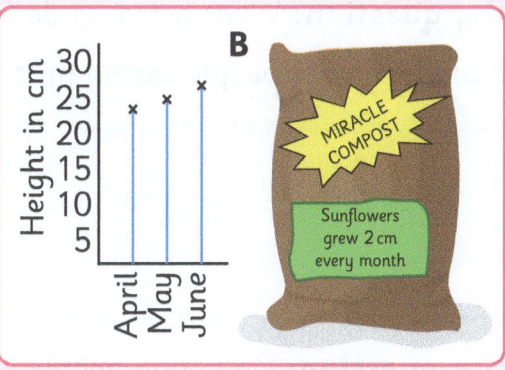

a) Which advert do you think might encourage more people to buy the compost? Explain your answer.

b) Which advert is more trustworthy? Explain your answer.

15.3 Understanding sampling bias

1 For each big question, say whether the sample is biased or unbiased. Explain your answer for each one.

a) **Big question:** What activities do young people enjoy doing in the summer holiday?

Sample: young people taking part in a musical theatre summer school experience.

b) **Big question:** Do young people in a school prefer using a laptop or a tablet when doing research in class?

Sample: every fifth name on the register for each class.

2 The students looked at each of these and decided that the samples are not good representations of the whole population. Why do you think they decided this? Explain your thinking.

	Big question	Sample	Explanation
a)	Which funfair ride is most popular?	People queuing for a rollercoaster ride at the funfair.	
b)	Which day are people most likely to go to a supermarket?	Shoppers in a supermarket at the weekend.	
c)	Which author is most popular with people aged 12 or under?	Adult customers in a bookshop.	

3 Suggest a sample that would be a good representation of the whole population for each big question in question 2.

	Big question	Unbiased sample
a)	Which funfair ride is most popular?	
b)	Which day are people most likely to go to a supermarket?	
c)	Which author is most popular with people aged 12 or under?	

⭐ **Challenge**

Some students read a newspaper article that interests them.

School traffic issues

Over the last 20 years, the number of students being driven to school has doubled. In some areas, one in every five cars on the road at 8·50 am is going to school.

They decide to inquire into how many students in their school are driven there regularly by car.

a) What might their big question be?

b) How might they collect data in a way that avoids bias?

1 There are 60 children in Primary 7. They have each been asked to choose one of these items of clothing as a leavers gift from the parent council: hoodie, polo shirt, rugby top, waterproof jacket.

Their choices are displayed in this pie chart.

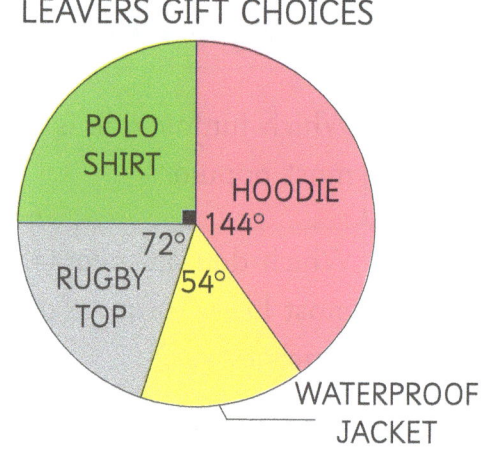

LEAVERS GIFT CHOICES

a) What was the least popular choice?

b) What percentage of the children chose a polo shirt?

c) How many children chose a hoodie?

★ Challenge

Riverside School and Parkway School took part in a sports festival. These pie charts show some information about the medals each school won.

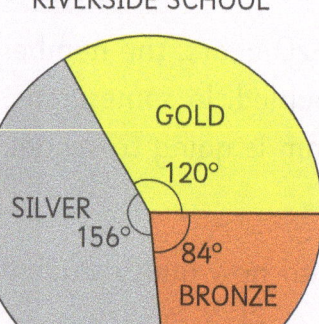

RIVERSIDE SCHOOL

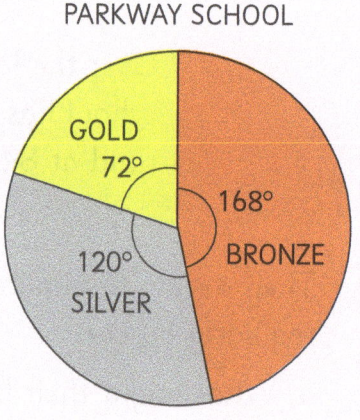

PARKWAY SCHOOL

a) If Riverside school won seven bronze medals, how many gold medals did they win?

b) A teacher says, "The pie charts show that Riverside School won more gold medals than Parkway School." Is this correct? Explain your thinking.

1 Some students have a set of 12 cards with numbers on them:

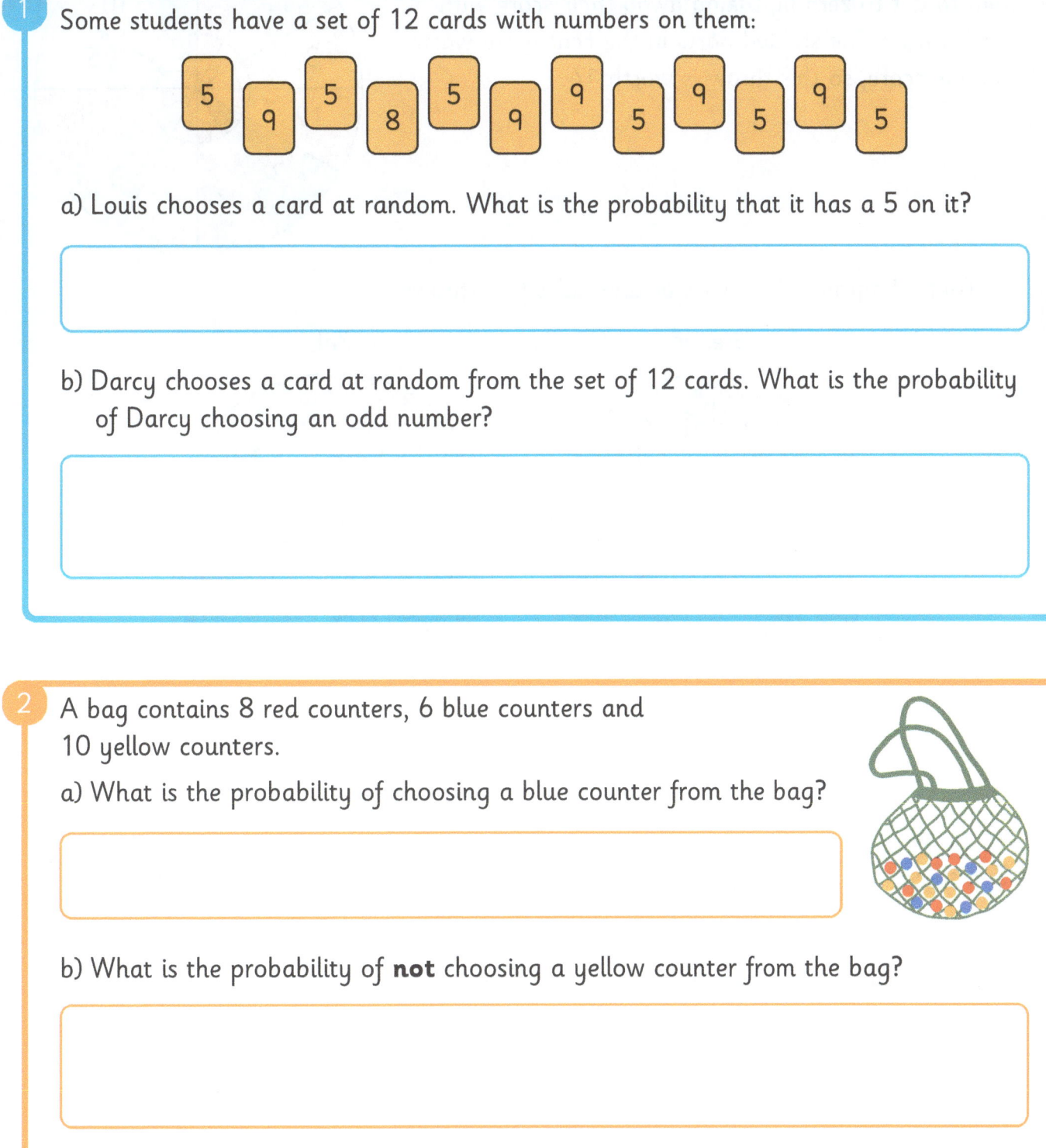

a) Louis chooses a card at random. What is the probability that it has a 5 on it?

b) Darcy chooses a card at random from the set of 12 cards. What is the probability of Darcy choosing an odd number?

2 A bag contains 8 red counters, 6 blue counters and 10 yellow counters.

a) What is the probability of choosing a blue counter from the bag?

b) What is the probability of **not** choosing a yellow counter from the bag?

c) 12 purple counters are added to the bag. What is the probability of picking a purple counter from the bag?

Gregor and Zak are playing a game by throwing a ball at this target. They start with 24 points and aim to get to zero by taking away their score with each throw. The shaded parts in the centre are worth double score, so this throw is worth 14.

a) These diagrams show Gregor and Zak's first throws.

Gregor

Zak

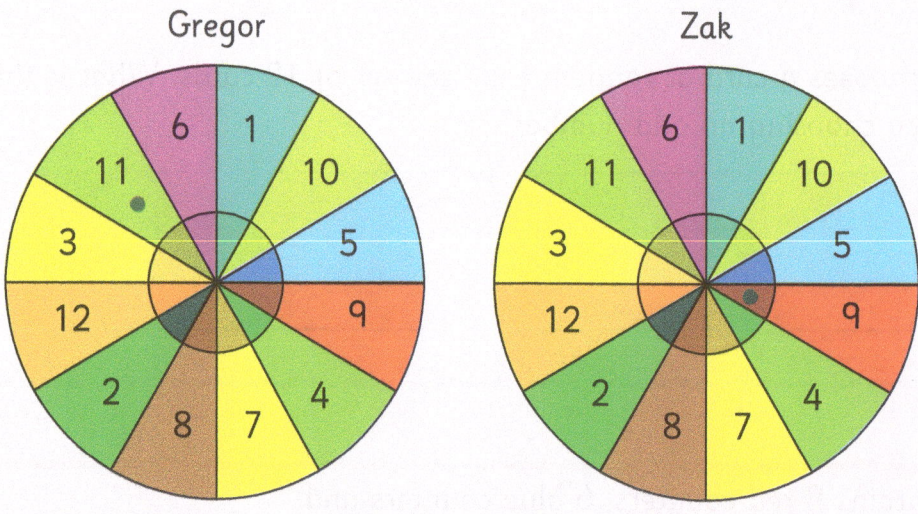

Gregor says: Oh no, I can't win with my next throw.

Zak says: I need to score 6. I can win with my next throw in two different ways.

What are the two different ways for Zak to score 6?

b) Which numbers have the highest probability of being scored with one throw? What is this probability?

001/16012025

10 9 8 7 6 5 4 3 2 1

ISBN 9780008741402

Published by
Leckie
An imprint of HarperCollins Publishers
Westerhill Road, Bishopbriggs, Glasgow, G64 2QT

T: 0844 576 8126 F: 0844 576 8131
leckiescotland@harpercollins.co.uk www.leckiescotland.co.uk

HarperCollins Publishers
Macken House, 39/40 Mayor Street Upper, Dublin 1, D01 C9W8, Ireland

Publisher: Fiona McGlade

Special thanks
Project editor: Peter Dennis
Layout: Jouve
Proofreader: Louise Robb

A CIP Catalogue record for this book is available from the British Library.

Acknowledgements
Images © Shutterstock.com

Printed in United Kingdom